LOFT
OFFICE
RESTAURANT
COMMERCIAL SPACE

LOFT 办公 餐饮 商业展示

高迪国际出版（香港）有限公司　编

君誉文化　策划

卢晓娟　崔倩　颜妍　李昕雨　刘若梦　赵世纯　译

大连理工大学出版社

图书在版编目(CIP)数据

LOFT：办公 餐饮 商业展示：汉英对照/高迪国际出版（香港）有限公司编；卢晓娟等译. —大连：大连理工大学出版社，2016.9
 ISBN 978-7-5685-0528-4

Ⅰ.①L… Ⅱ.①高…②卢… Ⅲ.①建筑设计—作品集—世界—现代 Ⅳ.①TU206

中国版本图书馆CIP数据核字（2016）第195710号

出版发行：大连理工大学出版社
　　　　　（地址：大连市软件园路80号　邮编：116023）
印　　刷：上海锦良印刷厂
幅面尺寸：235mm×310mm
印　　张：20
插　　页：4
出版时间：2016年9月第1版
印刷时间：2016年9月第1次印刷
责任编辑：张　泓
责任校对：王秀媛
封面设计：高迪国际
策　　划：君誉文化

ISBN 978-7-5685-0528-4
定　　价：368.00元

电　话：0411-84708842
传　真：0411-84701466
邮　购：0411-84708943
E-mail：designbookdutp@gmail.com
URL：http:// www.dutp.cn

如有质量问题请联系出版中心：（0411）84709043　84709246

PREFACE 1
序言一

Design of Renovation Office Space

The origin of loft style has its roots in the large, open spaces left derelict from unused factories and warehouses. At the turn of the 19th century, artists in Paris and later in New York City found the freedom of big space beneficial for the creation of their work.

Originally attractive as cheap studio space, the flexibility of tall, wide open loft space begins to inform the kind of work produced both in the scale of the art and the visual aesthetic. Raw loft space tends to be materially rich with exposed brick, heavy timber, riveted steel, or cast iron columns and light delivered from on high. Today's contemporary office environments value the same qualities that attracted artists to loft living more than a century ago: great natural light, expansive spaces prized for their spatial flexibility, and the visually stimulating quality of their interiors.

On many renovations, the most important decisions an architect makes are the selective decisions to do very little or nothing at all…The architect decides not to paint a rusty old ceiling, not to refinish battered old wood, and to salvage perfectly functioning, though tired looking equipment. Instead, the architect provides the necessary repairs to keep original machinery and retains the character of materials that have seen generations of human toil. As the digital revolution provides more striking visual effects in our everyday lives, seeing real materials and their authentic aging is profound.

It is true, the rapid evolution of technological change is providing new opportunities for collaboration and increasing efficiency of physical space. But we have also found that it is not the large format, high definition, technological marvel of the day that humans respond to. Initially, perhaps, but over time it is the rich patina inscribed from years of heavy human use on metal, wood and concrete that provide the visual backdrop for new creativity and innovation. It is the designer's responsibility to see through the clutter, resist the temptation to start wholly new, and instead retain the existing elements that add layers of history and create an authentic place.

We believe human beings have an inherent fascination with understanding history and their place in it. Renovating existing space for new uses provides us that opportunity in a very real and visceral way and we have a deep passion for refurbishing marginally viable buildings into new work places ready for their next chapter to be written.

Craig Norman and Agustin Enriquez V
Principals at GBD Architects Incorporated
克雷格·诺曼、奥古斯汀·安立奎·V
GBD建筑事务所创始人

改造办公空间的设计

Loft风格起源于废弃的工厂和仓库的宽阔的空间。在19世纪末20世纪初，巴黎和纽约的设计师们先后发现了大空间的自由度有利于他们的设计。

Loft空间最初作为一个廉价的工作室空间而吸引了设计师们，而后这高大宽敞的空间所具有的灵活性给了他们艺术规模和视觉上的美感。未经处理的loft空间因为有裸露的墙砖、厚重的木材、铆结钢或者铸铁立柱和从高处进来的光线营造出一种物质上的充实感。今天的现代办公环境和一个世纪前的艺术家在loft中所重视的是相同的品质，即充足的自然光、大空间引以为傲的空间灵活性和内部的视觉刺激。

在很多的室内翻修中，建筑师所做的最重要的决定是很少或甚至不做选择性的决策。比如，不要去粉刷一个生锈的旧天花板或是修补磨损的旧木头，挽救那些还完好运作但是看起来已经老旧的设备。相反，建筑师为原有的机械和特色材料提供必要的维护，因为从材料上可以看到几代人的辛苦劳作。虽然数字革命给我们的日常生活带来更强烈的视觉冲击效果，但看到真实的物质材料、感受它们的衰退更为重要。

的确，技术变革的快速发展提供了新的合作机会，并且提高了物理空间上的效率，但是同时我们也意识到人类所回应的不是当前时代的大规模的格式化、高清晰度和技术奇迹。或许最初是这样的，但是随着时间的推移，我们认识到是人类年复一年使用的金属、木头和混凝土所镌刻下的岁月痕迹为新的发明创新提供了视觉背景。设计师的责任就是从这些杂乱无章中找到条理，抑制住自己想要推翻从前重建一切的欲望，坚持保留已有的而且还在不断叠加的历史，并且创造一个真实的空间。

我们相信人类与生俱来就渴望理解历史和他们在历史中所处的位置。为了新的用途来改造现有空间的做法，以一种真实的、近乎直接的方式给了我们一个机会，使我们热衷于翻修可挽救的建筑，然后把它们变成一个可以书写新篇章的工作空间。

PREFACE 2
序言二

A loft is inherently a large adaptable open space. The trick is to keep true to this typology while creating functional and inspiring environments that fit the user's needs.

I will elaborate on challenges and qualities of the loft in office design, with Capco Bold Rocket as case study.

Our client relocated their offices to a 150-year-old 5 storeys steel and brick loft in Shoreditch, London – an area with a history of light industry.

The building's spirit is preserved by giving the brick walls and steel columns plenty of breathing room. These typical elements are made stronger: the brick walls are untouched and well lit; the steel columns and beams are black to stand out against light floors, walls and ceilings.

New inserts refer to the light industrial context. Raw or painted structural wood studs are exposed to create a visually stimulating scape. Without obstructing wide views, the open space is organized around volumes which float in the larger space, enhancing its loft character. Glass and timber partitions and doors maximize transparency.

The loft is a fitting typology for the bold program: except for the CEO office, there are no enclosed offices and no fixed desk spaces.

Each worker has a locker and uses the workplace like a club – they might have their favorite place to sit, but they select their own work setting every single day, picking a spot that best suits their moods and needs of the moment: collaborate, focus, present, and ideate.

Each floor has a variety of environments ranging from super flexible and informal to more structured; from intimate quiet booths to large meeting rooms. Variation within the open spaces is created by different seating arrangements: large communal tables; lounges with comfortable seating for impromptu informal meetings; "butcher blocks" – high boomerang shaped tables for a small group of people to stand and present, critique, and brainstorm.

Workers cycle intuitively through these different setups during the course of the day. While roaming around, there are more "serendipitous collisions" – informal chance encounters between workers of different departments that may spark innovative ideas.

The loft enables interesting juxtapositions without creating conflict: a pool table near the client meeting room; a large rocket sculpture over the staircase, informally drawing people downstairs; ping pong tables to let off steam adjacent to communal work tables; a cafe next to the reception; a podium floating between lounges.

The loft offers flexibility and adaptability. Furniture on wheels and versatile lightweight foam cubes allow for easy transformation of the open space. The podium functions as informal meeting place as well as seating for town hall presentations.

The loft office is not about squeezing in as many people as possible. It is about giving workers a broad range of experiences and interactions that keeps them energized and inspired. It is about transforming the office into a close knit community with happy and creative people, resulting in overall better performance. You don't own one office, you own the entire office.

Jeroen De Schrijver and Ellen Depoorter
Founders of D+DS Architecture Office
吉荣·德·斯赫雷弗、艾伦·德魄特
D+DS 建筑事务所创始人

　　Loft 是指可塑性极高的开放空间。其设计的要义在于保持其本色的同时创造出具有功能性和创造性的环境来满足使用者的需求。

　　我将用 Capco Bold Rocket 办公室的案例来详细阐述 loft 办公设计中的优势与难点。

　　我们的客户的新办公室位于一幢拥有一百五十年历史的五层钢砖结构厂房中。该建筑位于伦敦的肯迪奇区，这一地区以轻工业闻名。

　　通过给予砖墙和钢柱大量的呼吸空间，建筑的特点得以保留。而以下这些元素又得以增强：没有变化且有着良好的照明的砖墙；在浅色的地板、墙面和屋顶背景下格外显眼的黑色钢柱及横梁。

　　建筑内新的嵌入物也与其轻工业环境相适应。原木或涂漆的结构木立柱形成了视觉刺激。开放的空间围绕着漂浮在更大空间内的体量，打造了开阔的视野，更增强了其开敞的特点。木质和玻璃的隔断及屋门也最大限度地增强了空间的通透感。

　　空间设计与大胆的项目设计也完全符合：除总裁办公室外，楼内无封闭或固定的办公空间。

　　每个员工都有一个自己的储物柜并把工作场所当成俱乐部——员工可能有自己喜欢的位置，但是每天可以随自己的心意安排工作环境，选择最能满足自己当时的心情和需求的工作地点，无论这种需求是合作、专注、展示还是想象。

　　从超级灵活的、非正式的区域到正式场合，从安静的私人雅座到大型的会议室，每一楼层的空间设置都十分丰富。不同的开放空间设置源于不同的座椅安排，包括：大型的公用桌；为临时非正规会议准备的带有舒适座位的休息处；"实木桌"——类似于回飞镖形状的会议桌，方便小组成员站立、展示、评论和集体研讨。

　　在每日工作中，员工们都会经过这些不同的使用空间。但在漫步时，会有更多的"意外邂逅"——不同部门间员工的相遇可能会激发灵感。

　　Loft 空间使内部各部分并列设计却没有违和感：客户接待室设有台球桌；楼梯处矗立着一座大型火箭雕塑，吸引着下楼的人们的视线；乒乓球桌填补了公用办公桌间的空隙；接待处旁边就是咖啡室；休息室之间还设有展示台。

　　Loft 具有极强的适应性和灵活性。带有轮子的家具和多功能的轻型的泡沫隔间使得开放空间更容易进行转换。展示台不仅是临时的会议室，还是员工展示台。

　　Loft 办公室的意义不在于最大限度地容纳员工人数，而在于给予员工们广阔的体验和互动空间，以激发其活力与创造力；在于将办公室变为由快乐而充满创造力的人们组成的亲密社区，并以此提升公司整体绩效。你所拥有的不仅仅是一间办公室，而是整个办公空间。

PREFACE 3
序言三

Loft style revolves around notions of invention and reinvention. For a fast-growing city like Toronto, where Quadrangle was founded 30 years ago, this birth and rebirth is essential.

The city relies on the relevance of the existing building stock, as much as its new construction, to provide the spaces for an expanding population and workforce. Sites that were once occupied by the businesses of the past — garment and machine factories, for example — can be successfully transformed and updated to accommodate the city's present and future needs. Old industrial lofts are particularly valuable for this because their open plans and generous floor plates have a certain kind of flexibility that can be adapted into so many different purposes. The trick, though, is to preserve the architectural heritage and character of these structures, while enhancing their usability.

At Quadrangle, one of Toronto's largest architecture and interior design firms, we recently upgraded a 118-year-old former department store warehouse at 60 Atlantic Ave in the Liberty Village neighborhood, an area that boomed during the industrial revolution, and whose factories are now being turned into offices and condo developments. We strategically renovate and expand the building, embracing the collective memory of the city while transforming the outdated site into a dynamic hub. The interiors now have large open areas where people can work collaboratively. The outdoors act as a complement, where a new communal space allows for chance encounters and encourages the same collaboration and exchange of ideas taking place inside.

At the outset of the renovation, we begin by stripping away, revealing, reinforcing, power washing and sandblasting the building back to its original state. With the interiors, we clear out a series of renovations that have been done over the last century, which have left the spaces cluttered. To clarify the plan, we restore the brick and heavy timber beam structure to create spacious and welcoming interior environments that reflect the character of the old buildings while providing flexibility and functionality. Details including an artistic film with a historic map of the neighborhood that wraps the first floor windows, draw on the site's past to provide contemporary comforts to the users – in this case shade and privacy.

To enhance the connections to and through the building, a Corten steel and glass circulation core unify the building at all levels, improving the flow and access throughout (and all the circulation is universally accessible, ensuring that everyone, regardless of their abilities, can enjoy the spaces). The materiality boldly announces what is new and reflects the neighborhood's changing role from the industrial engine that helped build Toronto, into the home of an internationally connected, creative community.

To broaden the property's presence and value, we excavate the site, opening the basement level to a bright courtyard and outdoor beer garden that will draw patrons and create a vibrant property spanning the workday and well into the evening. Adding both hospitality and a sense of vibrancy to an area of the neighborhood currently lacking activity, we create a diverse program to enliven the building.

Adaptive reuse is, and continues to be, an opportunity for us to invent and reinvent. Our goal is activated buildings, and buildings that activate their environment – both inside and out. We strive to create spaces and places for people first and foremost. By ensuring that reinvention and invention of those spaces and places occur we make the kind of city, streets and neighborhoods that reflect the changing nature of urban and professional experiences.

Richard Witt
Principal of Quadrangle Architects
查理·韦特
Quadrangle 建筑事务所首席建筑师

Loft 风格的核心是创造与再创造。对于多伦多——30 年前 Quadrangle 建筑事务所成立于此——这样一个快速发展的城市而言，这种起源与复兴是必要的。

这座城市依靠现有建筑群间的联系以及新的建筑物，为增长的人口和扩张的劳动力提供空间。那些曾被旧有的商业模式——如服装厂和机械厂——占用的场地可以被成功转型和更新以适应城市现今以及未来的需求。旧工业 loft 空间在这方面尤其具有价值。它们的开放式规划和宽阔的楼面板具有一定的灵活性，可以通过改造而满足多种不同用途。然而，诀窍却在于如何在提高使用性的同时，使建筑遗产以及这些结构的特色得到保护。

Quadrangle 建筑事务所是多伦多规模最大的建筑和室内设计公司之一，我们最近对位于自由区大西洋大道 60 号的一家拥有 118 年历史的原百货公司仓库进行了改造。自由区在工业革命期间急速发展，那里的工厂现在都被改建为了办公区和公寓大厦。我们对这座建筑进行了策略性的翻修和扩建，保留了对于这座城市的共同回忆，同时把它从一个过时的建筑变为一个充满活力的动态中心。现在，建筑内部拥有巨大空间，人们可以在这里一起工作。户外空间作为补充空间，新的公共区域为人与人之间的偶遇提供了机会，并促进了室内相同的合作和思想交流的进行。

翻修之初，我们对建筑进行了剥离、暴露、加固、冲洗和喷沙，将其还原为原始状态。在建筑内部，我们对 20 世纪进行的一系列翻修进行了整理。那些翻修使空间显得杂乱。为使规划更明确，我们复原了砖和重木质横梁结构，在体现灵活性和功能性的同时，营造出可以反映这座古老建筑特征的宽敞的、热情的内部环境。翻修中注重了许多细节。其中包括我们用画着街区历史面貌的艺术薄膜将一层的窗户包覆起来，利用建筑场地的过去为使用者提供当代的舒适感。同时，薄膜也起到了遮阴和保护隐私的作用。

为了加强建筑与外部以及内部的联系，一个由考顿钢和玻璃建成的循环中心将建筑的各个平面联合在一起，提高了各方面的流动性和可接触性（所有的流通渠道都可以进入，保证了任何人，不管他们能力如何，都能享用这些空间）。大胆的选材向人们宣告了何为新事物，并展示了该街区从一个促进多伦多建设的工业引擎到一个与国际接轨、充满创造性的街区的角色上的转变。

为了扩展这座建筑的形象和价值，我们对其进行了开凿，将地下层改建为明亮的天井兼户外啤酒花园。它将吸引顾客前往，在白天和晚上都是一个充满活力的场所。对当前缺乏活力的街区，我们通过提升其中一个区域的好客感与活力创建了一个多样化的项目，使这座建筑重新散发出活力。

可适应再利用正在为并且将继续为我们进行创造和再创造提供机会。我们的目标是建造焕发活力的建筑以及由内而外激活其所处环境的建筑。我们努力奋斗的首要目标是为人们创建出空间和环境。通过确保空间和环境创造及再创造的完成，我们打造了能够反映城市及职业经历变化本质的城市、街道和街区。

CONTENTS
目录

012　Whole Foods Market Southern Pacific Regional Headquarters / 全食超市南太平洋地区总部
020　Actis Wunderman Office / 英联投资伟门办公室
030　Deluxe / Deluxe 公司
038　Donmar Dryden Street / 德莱顿唐马仓库剧院
048　Office Playtech / 游艺平台办公室
058　Red Bull / 红牛工作室
066　Capco Bold Rocket / Capco Bold Rocket 办公室
074　Heavybit / Heavybit 办公空间
084　Office and Showroom of Architectural Workshop Sergey Makhno / 谢尔盖·马克诺工作室办公室及陈列室
094　Red Bull Music Academy / 红牛音乐学院
104　SPARK Beijing Office / 思邦北京办公室
114　AKQA / AKQA 办公室
122　Comcast Silicon Valley Innovation Center / 康卡斯特硅谷创新中心
130　Corporate Office in Milan / 米兰分公司
138　Corus Quay / 哥鲁氏码头办公室
148　De Burgemeester / De Burgemeester 办公大楼
156　DRAFT Tokyo Office / 东京 DRAFT 办公室
164　Google Campus / 谷歌总部园区
174　Office Design of IND Architects Studio / IND 建筑师工作室办公室设计

182	Octapharma Brewery	/ Octapharma 啤酒厂
190	Office IMd Rotterdam	/ 鹿特丹 IMd 公司办公室
200	NeueHouse	/ 新空间
206	Yandex Stroganov Office	/ 斯特罗加诺夫 Yandex 办公区
214	Coca-Cola	/ 可口可乐办公室
222	Zendesk San Francisco Headquarters	/ Zendesk 旧金山总部
230	Valencia	/ 瓦伦西亚舞蹈厅
238	EHE Design Xiaohe Road Office	/ 易和设计小河路办公室
244	Walmart Brazil	/ 沃尔玛巴西办公室
252	Deliqatê Restaurant	/ Deliqatê 餐厅
260	Emporium Santa Isabel	/ 圣伊萨贝尔购物中心
266	Hunt a Lobster Restaurant	/ 猎寻龙虾餐厅
274	Il Giorno Canteen DM	/ 白日食堂设计手册
280	Korean Dessert Cafe Mu-A	/ 韩国 Mu-A 甜品咖啡馆
290	Murakami Restaurant	/ 村上日式餐厅
296	Factory 5	/ Factory 5 自行车商店
304	Russian Performance	/ 俄罗斯车库展览会
314	INDEX	/ 索引

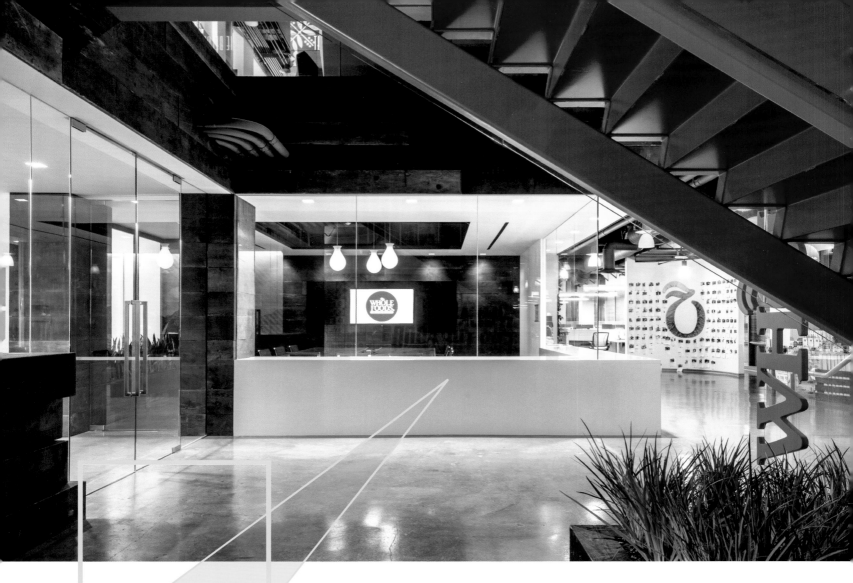

Whole Foods Market Southern Pacific Regional Headquarters

全食超市南太平洋地区总部

DESIGN COMPANY
Wirt Design Group
LOCATION
Glendale, California, U.S.A.
AREA
4,273 m²
PHOTOGRAPHER
Art Gray

Whole Foods Market recently took occupancy of its new Southern Pacific regional headquarters, designed by Los Angeles-based design firm, Wirt Design Group, located in Glendale, california. The new office space included two floors and outdoor space in a USGBC LEED Gold building.

The client wanted to create a space that would thoughtfully align with their company culture, mission, values, and brand. The goal was to develop a workplace that mirrored their retail environment where team members would live in the company culture of family, community, sustainability, health and transparency.

An open, interconnecting stair was critical to creating the sense of connectedness the client wanted to achieve in the space. A significant amount of glass was used throughout to emphasize open communication and transparency. With food as the core of their brand, the test kitchen – prominently located near the front entrance for maximum visibility – was the heart of the space.

Inspired by Whole Foods' 365 product branding, vibrant colors were strategically used for wayfinding and zone definition. Large-scale, nature-inspired graphics reflected the geographic regions the office serves.

The interior layout and furnishings made use of the building's full height glazing, bringing daylight and views of neighborhood and adjacent mountain ranges which, in turn, provided a physical and spiritual connection to the natural environment.

Because work happened anywhere, a wide range of alternative work spaces were designed beyond the workstations, including casual lounge areas with soft seating scattered through the open office, telephone rooms, quiet rooms and break room.

From the use of natural lighting to the reclaimed railcar siding, then to the exposed floors, the concepts of sustainability, re-use, local sourcing and minimalism tied the entire space together and aligned Whole Foods Market's culture to the workplace environment.

7th Floor Plan

8th Floor Furn plan

近来，全食超市进驻了其位于南太平洋地区的新总部，这个新总部由总部位于洛杉矶的设计公司沃特设计团队来进行设计。该公司位于加利福利亚州格伦代尔市。新办公地点位于一座获得美国绿色建筑委员会LEED金奖大楼，占据了两层楼和室外空间。

业主希望缔造出一个非常切合该公司文化、宗旨、价值观和品牌的空间。设计目的是要打造出一个能反映其零售环境的工作场所，在那里，团队成员可以生活在一个集家庭、社区、持续性、健康与透明性于一体的公司文化之中。

一座敞开式的互连楼梯至关重要，它能打造出业主力求的那种空间之内的连通感。设计大面积地使用玻璃来凸显开放式的沟通交流和透明度的特点。食品作为品牌经营的核心内容，实验厨房——离前门入口最近，实现最高的能见度——是空间的中心地带。

受到全食行业365种产品品牌化的启发，设计策略地使用鲜亮的颜色来做道路标识和区域划分。大型的、自然风格的平面图案还能反映办公室所服务的区域。

室内布局和陈设利用建筑物的落地玻璃窗进行采光、观赏四周风景和不远处绵延起伏的山脉，搭建起与大自然全身心接触的桥梁。

由于任何地方都有可能碰到工作问题，所以，不止局限于固定的工作地点，设计还增设了各种各样的另外的工作场所，包括配有软椅的遍布开放式办公区的休闲休息区、电话间、静音房和休息室。

从使用自然采光到回收车厢壁板再到裸露的地面，可持续性、再利用、地方采购和极简主义的观念把整个空间紧紧地联系在一起，也使全食超市文化与其工作环境相吻合。

Actis Wunderman Office

英联投资伟门办公室

DESIGNER
Nikolay Milovidov, Anna Dykhanova, Sergio Markin, Tatiana Kulikova
DESIGN COMPANY
UNK project
LOCATION
Moscow, Russia
AREA
1,400 m²
PHOTOGRAPHER
Sergio Anan'ev

Actis Wunderman is the leader in the field of digital-marketing in the Russian market. The desire to create an office which fully opens an independent way of thinking is the cornerstone of idea of the organization of the office space. For this purpose the platform with the special atmosphere in "Arma" – the quarter of creativity, is chosen.

The new office represents an open space with accurate zoning – from the most silent department (with offices for directors and accountant department) to the noisiest and the most creative department through an account-zone and guest lounge. The customer initially sets the accurate task to create a separate client zone from where guests mustn't get to office. Thus in the plan a closed lounge with two glass doors (one of them is a false door) through which the customer can peep at working process without distracting employees appeared. Besides, glass partitions are additional light sources.

The separate entrance for workers which opens directly into a working zone is made. Important aspects in open space are comfortable acoustic conditions. Wherefore architects tries to divide zones of the room with big touches for meeting rooms and offices, and light furniture designs, slate boards for creation of cozier space. As "traps" of sound the punched information "cloudlets" over workplaces are used. At windows sound-absorbing panels are also provided. Everything in the office is focused on creation of comfortable working conditions.

No less important factor is the light decision. The customer at once pays attention to that employees most often don't use the general lighting (they like to work in the semi-darkened space). On each workplace, also according to sketches of bureau, the own light source which resolves an issue of comfort is an important element of a decor. But, nevertheless, classical "technical light" is installed on a ceiling.

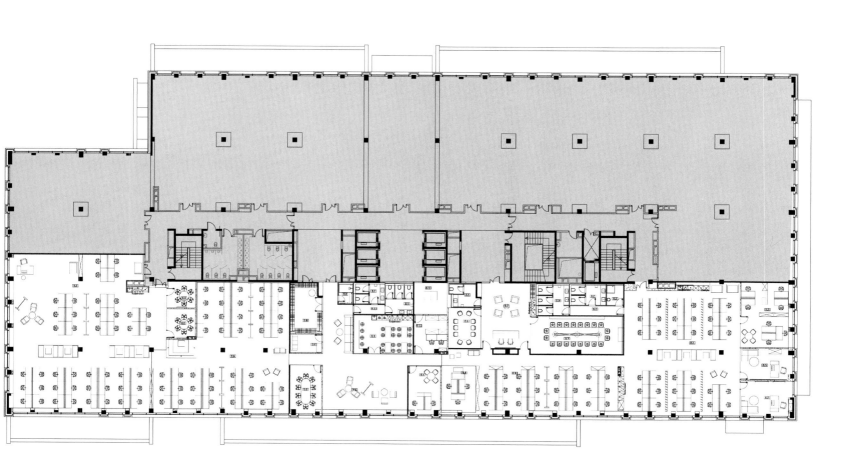

英联投资伟门公司是俄罗斯市场数字营销这一领域的领头羊。意在打造一个能完全开启独立思维方式的办公室是其办公空间组织理念的基石。为此，其选择了充满创意的"Arma"这种特殊氛围的平台。

新办公室呈现出一个分区明确的开放空间——通过一个开户区和宾客休息室可以从最安静的部门（含董事、会计部门办公室）到最喧嚣、最具创意的部门。客户最初提出过明确的任务来打造一个独立客户区，从那里客人们不必进入办公室。因此，该计划设计了一个有两扇玻璃门的封闭休息室（其中一扇为假门），透过门，客户可以观看工作流程而不打扰到员工。此外，玻璃隔板提供了额外光源。

员工专用的单独入口可以直达工作区域。开放空间的重要方面在于舒适的听觉环境。因此，建筑师试着把空间划分出会议室和办公室，并使用轻便家具设计和石板来缔造出更为舒适的空间。为达到隔音效果，冲孔材料板犹如轻盈的"彩云"在办公场所加以使用。窗口同样也设有吸音面板。办公室的一切设计都旨在打造一种舒适的办公环境。

光线选择也很重要。客户曾经注意到，员工往往不使用普通照明（他们喜欢在有点暗的地方工作）。设计也根据办公室的设计图纸，在每一个工作场所单独提供光源，既解决了舒适问题也成为重要的装饰元素。但是，尽管如此，传统的"技术光源"还是会安装在天花板上。

Deluxe

Deluxe 公司

DESIGN COMPANY
Quadrangle Architects

PROJECT TEAM
Ted Shore, David Takacs, Caroline Robbie, Rob Dyson, Julie Mroczkowski, Panyada Wangpongpipat

CONSULTANT
Integral Group

MECHANICAL ENGINEERING
Jablonsky Ast & Partners (Electrical), Pilchner Schoustal International (Structural)

CLIENT
Deluxe Entertainment

LOCATION
Toronto, Ontario, Canada

AREA
5,750 m²

PHOTOGRAPHER
Bob Gundu

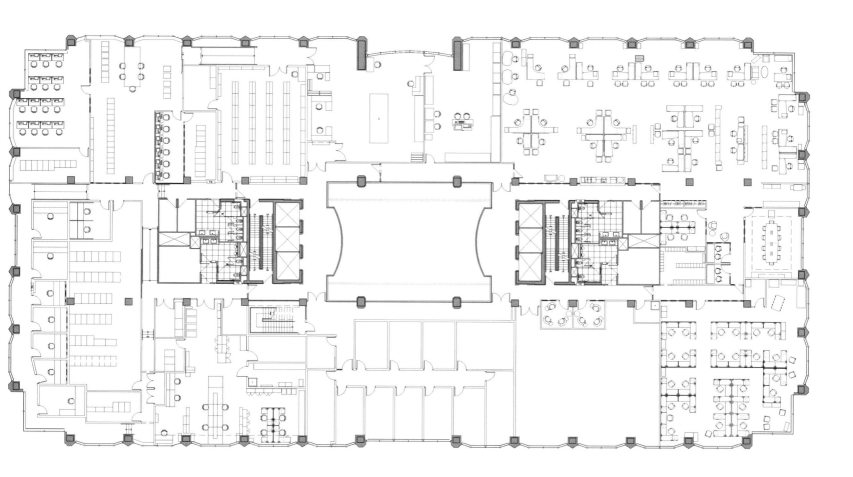

Deluxe is globally renowned for its media services including digital post-production, asset management and copyright protection for a diverse and global client list. The Toronto facility relocated to the top three floors of an existing 1990s data center building which was retrofitted to meet their highly technical and client-service oriented needs.

Quadrangle was engaged by Deluxe on two fronts: firstly, to provide pre-lease services to test a variety of potential sites, and secondly to leverage its industry knowledge in understanding work flow, environmental considerations, the detailed technical and audio requirements of high-end production facilities and the complex information technology infrastructure required to support the operations of a progressive content media environment.

Working closely with the client through design charrettes to understand the Deluxe brand, Quadrangle is able to develop the space as an extension of Deluxe's unique culture, reflecting the diversity of their business offering and commitment to client care. The facility includes theaters in a variety of sizes and sophistication, production suites, mixing stages, open office work areas and data rooms, as well as a large communal cafeteria lounge with a VIP dining room. Large areas of glass promote an open concept space that is unusual in technology and production environments.

The design and space plan gives prominence to key areas such as theaters, sound stages and quality control rooms. In order to accommodate two cinema-experience mix theaters, Quadrangle adds a sufficient volume to the roof of the building to create two theaters with 7.3 m clear ceiling height. The design is particularly successful in the acoustical separation of noise production and noise-sensitive functional areas.

Deluxe at 901 King West gives Toronto a post-production facility that rivals any in the world.

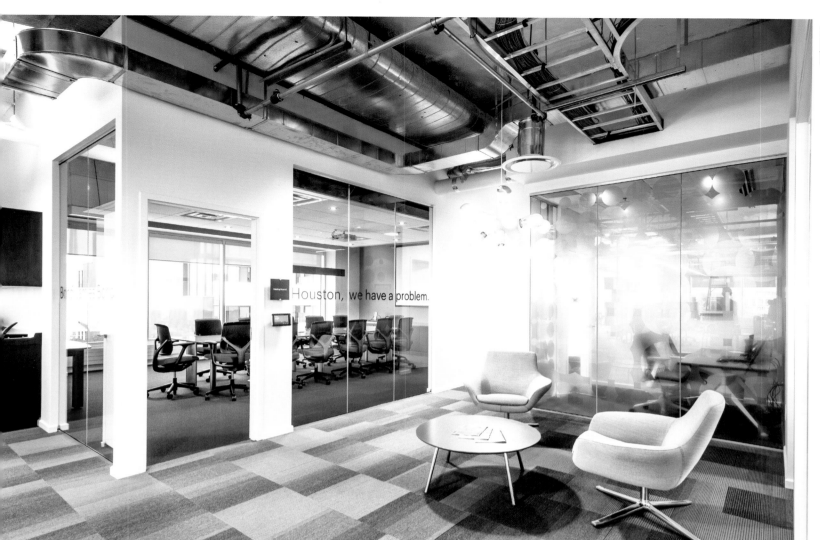

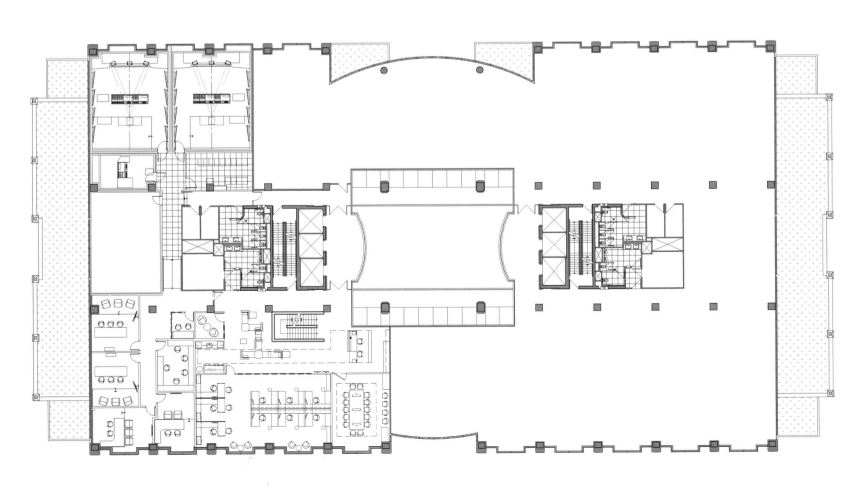

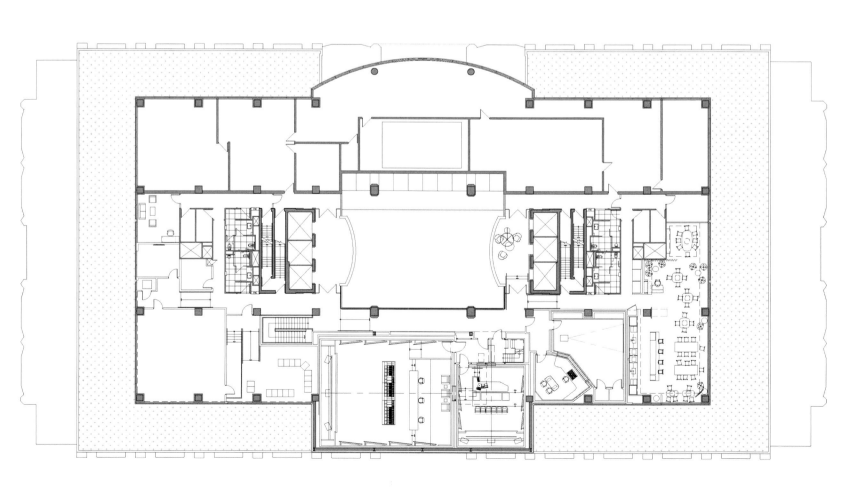

Deluxe 公司拥有全球多样化的客户群，媒体服务全球知名，服务包括数字化后期制作、资产管理和后期版权保护。其位于多伦多的公司基地搬迁至现有的一座 20 世纪 90 年代的数据中心大楼的最高三层。为了满足公司较高的技术和客户服务导向的需求，公司对这座大楼进行了翻新。

Quadrangle 建筑事务所受聘于 Deluxe 公司进行两个方面的改造。一方面，提供预租服务以达到测试各种潜在场所的目的；另一方面，充分利用行业知识，了解工作流程，考虑环境因素，满足公司对高端制作设备细致的技术和音频要求，以及通过使用复杂的信息技术基础设施支持正在蓬勃发展的深度内容媒体大环境所需的的运转和操作。

通过研讨会与客户紧密合作，了解了 Deluxe 公司的品牌，Quadrangle 建筑事务所将开发空间做为 Deluxe 公司的独特文化的延伸，展现该公司业务服务的多样性和对客户的承诺和关怀。楼内设施包括不同大小、设计复杂的剧场、还有制作室、混合舞台、开放式办公区域和信息室以及一个带有贵宾餐厅的大型公共餐厅。大面积玻璃幕墙的设计在技术生产环境中与众不同，向人们展现的是一种开放式的空间理念。

设计和空间规划注重突出剧场、摄影棚和质量监管室等重点区域。为了容纳两个影院级体验的混合剧场，Quadrangle 建筑事务所给建筑添加了一个巨大的屋顶，使两个剧场有 7.3 m 净高。建筑团队对噪声产生区和噪声敏感区的声音分离设计尤为成功。

位于国王西街 901 号的 Deluxe 公司让多伦多在后期制作设施上可以与世界匹敌。

Donmar Dryden Street

德莱顿唐马仓库剧院

DESIGN COMPANY
Haworth Tompkins Architects
PROJECT MANAGER
Vincent Wang
ENGINEERING
Price & Myers (Structural), Skelly & Couch (Service)
CONSULTANT
Charcoalblue (Theater), DP9 (Planning), Gillieron Scott Acoustic Design (Acoustic)

LOCATION
London, UK
AREA
863 m²
SCENOGRAPHER
Lucy Osborne
PHOTOGRAPHER
Philip Vile

The Donmar Warehouse is a leading producing theater company located in London's Covent Garden. Having acquired a small 19th Century warehouse building nearby Dryden Street, the theater asks Haworth Tompkins Architects to convert it for rehearsal, education and support facilities. The challenge is to design a convivial and creative professional working environment within the constraints of a limited budget and a tightly enclosed site.

By removing existing floors, extending the roof space and re-planning the circulation routes, the new facility yields a double height rehearsal room of a similar size to the main Donmar stage, a street fronting green room, administration offices, a large education studio in the roof rafters and a new attic apartment for visiting artists.

Stripped back and partially demolished walls and ceilings have been left raw as a suitably theatrical backdrop to the Donmar's working life, with a new polychromatic staircase, hand painted by Haworth Tompkins' regular collaborating artist Antoni Malinowski, as the warm heart of the building. Scenographer and production designer Lucy Osborne has collaborated closely on finishing and furnishings, reinforcing the collective sense of a theater production as much as a construction project.

As for many of Haworth Tompkins' creative working spaces in existing buildings, the personality of the historic architecture has been allowed to set the tone, with a provisional, loose-fit language of new additions setting up a fluid, adaptable relationship of new and old. Materials added have been simple and straightforward – whitewashed plywood, painted timber beams and wall coverings, waxed mild steel – to complement the richly patinated texture of the found surfaces. The aim is very much for a benign occupation rather than an obliteration of the original fabric.

Steve Tompkins, Director of Haworth Tompkins said: "the project has been a joint exercise in wringing the maximum creative potential from an ordinary London building on a tight budget. Working with Josie Rourke and her team has been great fun and, once again, our ongoing collaboration with artist Antoni Malinowski is central to the design."

SITE LOCATION PLAN
DONMAR DRYDEN STREET

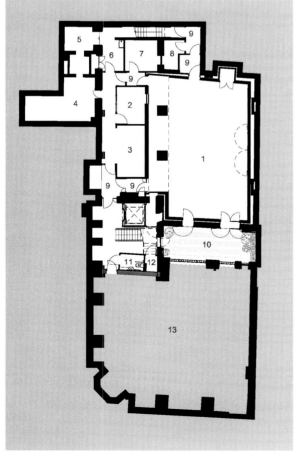

BASEMENT PLAN

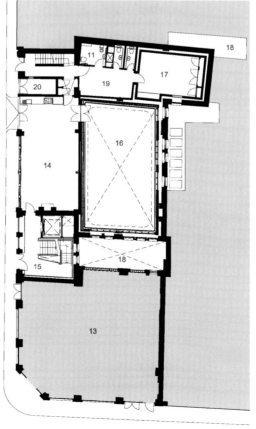

DRYDEN STREET

ARNE STREET

GROUND FLOOR PLAN

ANNOTATION KEY

1 REHEARSAL ROOM
2 STAGE MANAGEMENT
3 DRESSING / SUPPORT AREA
4 PROP STORE / MAKING SPACE
5 STORE
6 SUPPORT SPACE / CLEANERS CUPBOARD
7 BOILER ROOM
8 ELECTRIC INTAKE ROOM
9 LOBBY
10 LIGHTWELL
11 ACCESSIBLE W.C.
12 SERVER ROOM
13 NO. 5 DRYDEN STREET
14 GREEN ROOM
15 STAIR
16 VOID OVER REHEARSAL ROOM
17 MUSIC PRACTICE ROOM
18 EXTERNAL VOID
19 CYCLE STORE
20 BIN STORE
21 MAIN OFFICE SPACE
22 INFORMAL MEETING AREA
23 CASTING ROOM
24 EXECUTIVE OFFICE
25 MEETING ROOM
26 PAINTING / STATIONARY CUPBOARD
27 LIBRARY / ARCHIVE / QUIET ROOM
28 EDUCATION STUDIO
29 CASTING / AUDITIONS / MEETING ROOM
30 LIVING ROOM
31 BEDROOM
32 BATHROOM

N

FIRST FLOOR PLAN

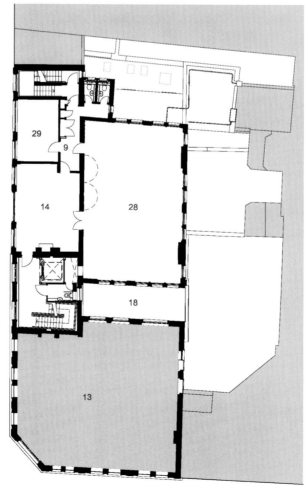

SECOND FLOOR PLAN

ANNOTATION KEY

1. REHEARSAL ROOM
2. STAGE MANAGEMENT
3. DRESSING / SUPPORT AREA
4. PROP STORE / MAKING SPACE
5. STORE
6. SUPPORT SPACE / CLEANERS CUPBOARD
7. BOILER ROOM
8. ELECTRIC INTAKE ROOM
9. LOBBY
10. LIGHTWELL
11. ACCESSIBLE W.C.
12. SERVER ROOM
13. NO. 5 DRYDEN STREET
14. GREEN ROOM
15. STAIR
16. VOID OVER REHEARSAL ROOM
17. MUSIC PRACTICE ROOM
18. EXTERNAL VOID
19. CYCLE STORE
20. BIN STORE
21. MAIN OFFICE SPACE
22. INFORMAL MEETING AREA
23. CASTING ROOM
24. EXECUTIVE OFFICE
25. MEETING ROOM
26. PAINTING / STATIONARY CUPBOARD
27. LIBRARY / ARCHIVE / QUIET ROOM
28. EDUCATION STUDIO
29. CASTING / AUDITIONS / MEETING ROOM
30. LIVING ROOM
31. BEDROOM
32. BATHROOM

  唐马仓库剧院是位于伦敦科文特花园的一家领先的戏剧制作公司。在购得附近德莱顿街上一家19世纪的小型仓库建筑后，剧院让霍沃思·汤普金斯建筑公司将其改建成用于排演、教育和拥有配套设施的场地。设计的难题是有限的预算以及要在一块四周被紧密围绕的场地内设计一个既轻松欢乐、又有创意的专业化工作环境。

  通过移除现有的楼层，延伸屋顶空间，重新设计室内流通线，新的建筑被规划成拥有一个双倍层高的和唐马仓库剧院主戏台一样大小的排演室，一间面向街道的演员休息室和管理办公室，另一间位于屋顶横梁上的大型教育工作室和给客座艺人居住的全新阁楼公寓。

  而作为唐马仓库剧院运作生涯中恰如其分的戏剧背景，脱落的墙皮、部分被毁掉的墙壁和天花板则保留着其原始的风貌未作修缮。由霍沃思·汤普金斯建筑公司定期合作的艺术家安东尼·马利诺夫斯基手工上漆的多色楼梯成为整个建筑的中心。舞台美学师和制作设计师露西·奥斯本密切配合项目修整和布置房间，如同建筑项目一样增强了整个剧院的氛围。

  霍沃思·汤普金斯建筑公司在现有建筑中设计的许多富有创意的工作空间，使这栋历史建筑的个性特点确立了建筑的基调，增添新的设计元素配以飘渺的、松散的表现方式在新旧之间建立了一种流动的、更有适应力的关系。新增的材料简单明确——白色涂料粉刷过的胶合板，上了漆的木横梁和墙面材料，打了蜡的低碳钢——用于补充建筑表面覆满绿锈的质地。设计目标更像是对建筑物的良性占用，而非消除原有建筑结构。

  霍沃思·汤普金斯建筑公司总监史蒂夫·汤普金斯表示："这个项目是在紧张的预算下，在一栋伦敦的普通建筑里发挥最大创意潜能的联合设计活动。与乔茜·洛克以及她的团队一起工作很有乐趣，与艺术家安东尼·马利诺夫斯基的再次合作对设计的完整性更是极为重要。"

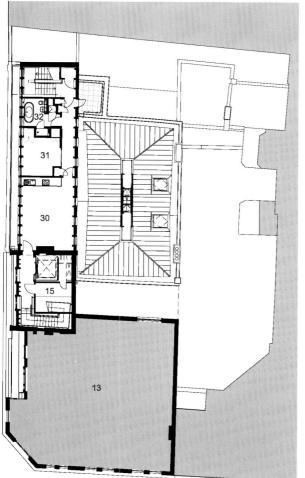

THIRD FLOOR PLAN

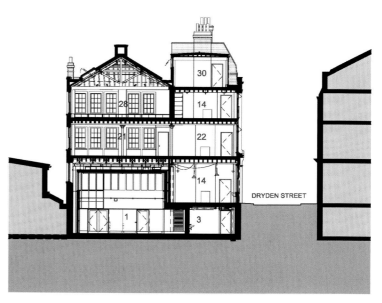

SECTION

ANNOTATION KEY

1	REHEARSAL ROOM	21	MAIN OFFICE SPACE
2	STAGE MANAGEMENT	22	INFORMAL MEETING AREA
3	DRESSING / SUPPORT AREA	23	CASTING ROOM
4	PROP STORE / MAKING SPACE	24	EXECUTIVE OFFICE
5	STORE	25	MEETING ROOM
6	SUPPORT SPACE / CLEANERS CUPBOARD	26	PAINTING / STATIONARY CUPBOARD
7	BOILER ROOM	27	LIBRARY / ARCHIVE / QUIET ROOM
8	ELECTRIC INTAKE ROOM	28	EDUCATION STUDIO
9	LOBBY	29	CASTING / AUDITIONS / MEETING ROOM
10	LIGHTWELL	30	LIVING ROOM
11	ACCESSIBLE W.C.	31	BEDROOM
12	SERVER ROOM	32	BATHROOM
13	NO. 5 DRYDEN STREET		
14	GREEN ROOM		
15	STAIR		
16	VOID OVER REHEARSAL ROOM		
17	MUSIC PRACTICE ROOM		
18	EXTERNAL VOID		
19	CYCLE STORE		
20	BIN STORE		

Office Playtech

游艺平台办公室

DESIGN COMPANY
Soesthetic Group
LOCATION
Kiev, Ukraine
AREA
3,000 m²
PHOTOGRAPHER
Alex Pedko

The office is located in the new business center Gulliver. A major feature is the panoramic view along the entire perimeter of the building. To incorporate this, the designers are trying to make the interior with maximum light, transparent and cosy at the same time.

To maximize the ceiling height, the designers leave the concrete ceiling slabs open with visible engineering structures.

Given it is a large office, the designers focus their attention on highlighting the public use areas of the reception, coffee points, relaxation and activity areas.

The main element of the activity area becomes the GAME BOX. It's made of perforated MDF with inner surfaces of black tempered glass. Screens for playstation games are installed behind the glass. A LED profile runs along the perimeter. It reflects on the glass surfaces and creates an unusual virtual atmosphere. All features connect at an accurate 45 degrees to make the complex technological interactions work.

The reception is made of roast metal with gradient perforated pattern. Lighting hidden behind the perforation creates a pattern on the floor. The shape of the perforation is the same as the cutting on the GAME BOX facades and includes the Playtech logo detail. The designers pay attention to the visual graphical navigation, so it lets the company employees understand the floor layouts and ease the use of public areas and subjects.

План с расстановкой мебели 21 эт.
М 1:100

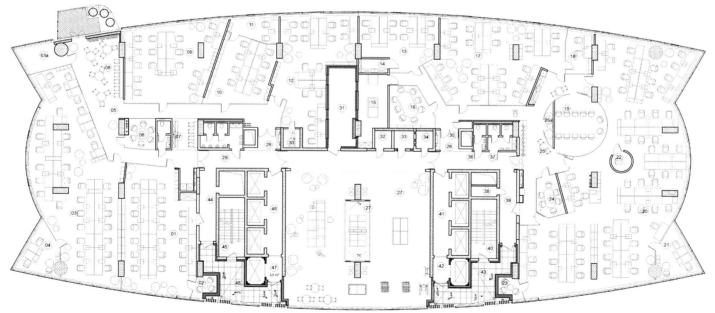

Экспликация помещений

№	Наименование	Площадь	№	Наименование	Площадь	№	Наименование	Площадь	№	Наименование	Площадь
2101	Офисное помещние	91.3	2114	Кладовая	4.4	2226	Противопожарный Коридор	22.4	2139	Противопожарный Коридор	13.1
2102	Переговорная	4.4	2115	Телекомуникационная	19.2	2127	Зона отдыха	232.1	2140	Лестничная клетка	15.9
2103	Офисное помещние	147.5	2116	Переговорная	17.1	2128	Противопожарный Коридор	22.4	2141	Лифтовый холл	13.9
2103а	Комната отдыха	11.6	2117	Офисное помещение	100.5	2129	Мужской санузел	9.2	2142	Лифтовый холл	3.9
2104	Кабинет	14.7	2118	Кабинет	11.0	2130	Помещение ОВ	1.5	2143	Балкон	-
2105	Коридор	84.6	2119	Переговорная	29.1	2131	Венткамера	22.4	2144	Противопожарный Коридор	13.1
2106	Переговорная	9.8	2120	Офисное помещение	275.1	2132	Электрощитовая	3.1	2145	Лестничная клетка	15.8
2107	Санузел унисекс	3.8	2221	Кабинет	14.6	2133	Электрощитовая	3.7	2146	Лифтовый холл	13.9
2108	Кафе - переговорная	16.2	2222	Капсула звукозаписи	3.4	2134	Электрощитовая	3.1	2147	Лифтовый холл	3.9
2109	Офисное помещние	74.2	2223	Переговорная	13.6	2135	Кладовая	0.9	2148	Балкон	-
2110	Офисное помещние	40.6	2224	Переговорная	13.6	2136	Санузел унисекс	3.2			
2111	Кабинет	11.6	2225	Коридор	74.7	2137	Женский санузел	6.4			
2112	Офисное помещние	76.7	25а	Кафе - переговорная	3.3	2138	Помещение СС	2.8			
2113	Офисное помещние	39.7									

Общая	1619.8
Полезная	1404.8

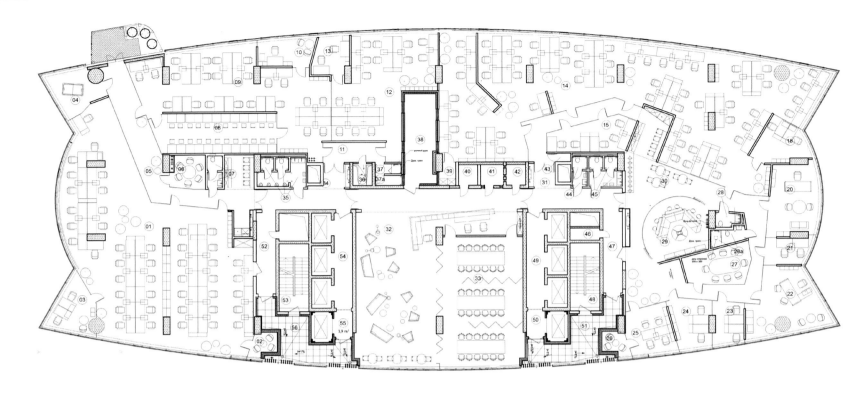

План с расстановкой мебели 22 эт.
М 1:100

Экспликация помещений

2201	Офисное помещение	243.2	2215	Коридор	5.5	2229	Переговорная	27.3	
2202	Переговорная	4.4	2216	Офисное помещение	78.9	2230	Коридор	112.4	
2203	Кабинет	14.7	2217	Переговорная	9.0	2231	Противопожарный коридор	22.4	
2204	Комната отдыха	13.7	2218	Кабинет	19.1	2232	Рецепция, зона отдыха	143.2	
2205	Коридор	55.5	2219	Кабинет	14.7	2233	Конференц-зал трансформер	87.6	
2206	Переговорная	9.4	2220	Кабинет	27.1	2234	Противопожарный коридор	22.4	
2207	Санузел унисекс	3.8	2221	Кабинет	9.2	2235	Мужской санузел	8.8	
2208	Зал игровых терминалов	54.6	2222	Кабинет	33.2	2236	Помещение ОВ	1.5	
2209	Офисное помещение	101.3	2223	Кабинет	10.1	2237 / 2237а	Душевая / Кладовая	4.6	
2210	Кабинет	11.8	2224	Директорская, Бухгалтерия	15.0	2238	Венткамера	22.4	
2211	Коридор	9.9	2225	Кабинет	16.7	2239	Кладовая	2.6	
2212	Офисное помещение	72.1	2226	Переговорная	4.5	2240	Электрощитовая	3.1	
2213	Кабинет	10.7	2227	Переговорная	17.9	2241	Электрощитовая	3.7	
2214	Офисное помещение	213.1	2228 / 2228а	Санузел унисекс и С/У администрации	2.8 / 4.6	2242	Электрощитовая	3.1	
2243	Кладовая	0.9							
2244	Санузел унисекс	3.6							
2245	Женский санузел	6.1							
2246	Помщение СС	2.8							
2247	Противопожарный коридор	13.1							
2248	Лестничная клетка	15.9							
2249	Лифтовый холл	13.9							
2250	Лифтовый холл	3.9							
2251	Балкон	-							
2252	Противопожарный коридор	13.1							
2253	Лестничная клетка	15.8							
2254	Лифтовый холл	13.9							
2255	Лифтовый холл	3.9							
2256	Балкон	-							

Общая 1521.8
Полезная 1399.7

这间办公室地处新兴商业中心格列佛，最大的特点就是沿着建筑四周可饱览周边风光。为体现这一特点，设计师们最大程度地利用内部光线，同时使室内空间具有透明、舒适的效果。

为最大化室内净高，设计师们采用了开放式的天花板设计，因此建筑内部混凝土顶棚的工程结构清晰可见。

因为办公室的面积相当大，所以设计师们重点打造了接待区、咖啡角及休闲娱乐区这些公共区域。

活动区最重要的设计元素是"游戏盒子"。"游戏盒子"外部由带有孔洞的中密度纤维板制成，内部表面则为黑色的钢化玻璃，游戏机的屏幕就安装在玻璃之后。"盒子"边缘饰有LED，其光线反射在玻璃上，呈现出梦幻的视觉效果，所有这些都以精准的45度角连接在一起使复杂的技术互动运作起来。

接待处由烤制金属制成，上面带有倾斜的孔洞图案，隐藏在后面的灯光透过这些孔洞在地面形成图形。孔洞的形状同"游戏盒子"正面的切割形状完全一致，也包括游艺科技标志的细节。设计师们还特别设计了视觉图形导向，以便于公司员工了解楼层布局，方便他们使用这些公共区域和设施。

NEW YORK

PLEASE PUT ALL
MILK PACKS

BACK TO THE
FRIDGE!

GUYS,
EACH FRIDAY

IN THE END OF THE WORKING DAY
ALL SHORT LIVED COMMODITIES
WILL BE **THROWN AWAY** –
INCLUDING FOOD IN
**YOUR LUNCH
BOXES.**

Red Bull

红牛工作室

DESIGN COMPANY
Giant Leap
LOCATION
Cape Town, South Africa
AREA
1,600 m²
PHOTOGRAPHER
Adam Leach

Giant Leap Workspace Specialists was commissioned by the highest selling energy drink in the world, Red Bull, to create the interior space of their new South African headoffice in Cape Town.

Importance was placed on understanding the company's brand and culture. Red Bull wanted creative, yet functional space where their staff could work individually as well as collaborate as a team. To achieve this, Giant Leap learnt all they could about Red Bull's culture and what their goals with their new two-storey offices were. They studied their staff and how they worked, how much time they spent in the office, and what tasks they most often performed.

The 1,600 m² space was designed around their people and their brand. The workspace specialists introduced walls showcasing their incredible action-packed adventurers, a racing games room for staff and war rooms for strategies. The brief was to deliver a high-end, yet funky space to represent their standing as a global brand. It also needed to fit within a budget.

The space was designed for around 74 staff, allowing for growth of up to 15 people. Technology was an integral part of the design brief since Red Bull communicated globally as well as with their Johannesburg offices. Giant Leap designed the space around an industrial look with no ceilings and all the services exposed, taking into account that the offices needed to be acoustically sound proofed. The floor space was populated over two floors in a way that encouraged staff to move between the floors while eliminating any hierarchical structures.

The kitchen area was designed for lunches and doubles as an area that could be transformed for entertaining, in true Red Bull spirit.

It was important to work with the natural light and to let it flow through the office environment. The final product was a space that allowed for collaboration and privacy to co-exist in a noise-controlled environment. The modern, high-energy design showed off their brand, which was visible throughout.

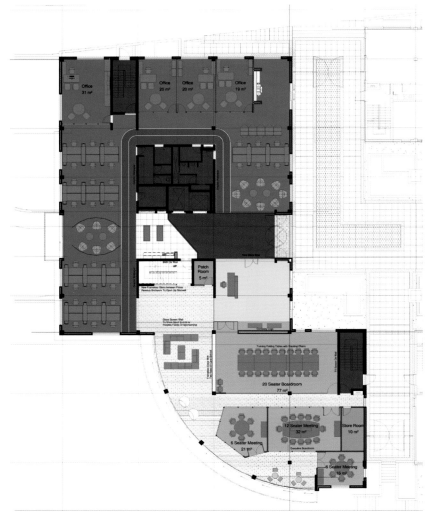

① D3.100 Ground Floor General Layout
1:100

② D3.100 Ground Floor General Layout
1:200

飞跃设计事务所的工作空间设计专家受拥有世界最大销量的功能饮料公司——红牛的委托，设计了位于开普敦新的红牛南非公司总部的内部空间。

此项目的重中之重是要理解这家公司的品牌与文化。红牛公司想要给职员们提供一个既创新又实用的空间，员工既可以独立工作，又可以与人合作。为了达成此目标，飞跃设计事务所竭尽所能去学习红牛公司的企业文化，努力了解公司对于这个有着两层高的新办公室的期望。他们调查了红牛职员的工作方式，在办公室的时间，以及经常做的工作类型。

这个 1 600 m² 的空间是围绕着红牛职员和品牌而设计的。空间设计专家设计了公司创立史展示墙，上面介绍了公司充满活力的员工们；还有一间为职员设计的游戏房和公司制定战略的企划室。总而言之，这间办公室要营造出高端又时髦的氛围，还要体现出全球大品牌的风范。当然，它的花费也需要控制在预算之内。

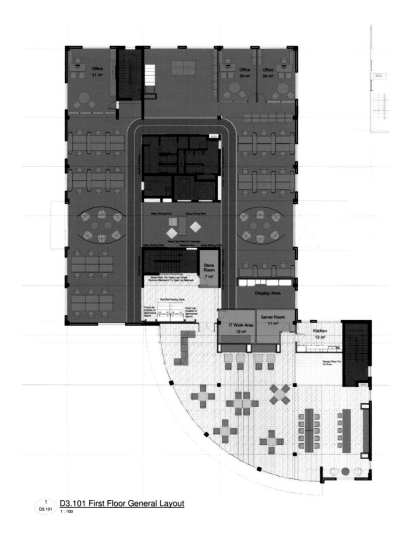

① D3.101 First Floor General Layout
1:100

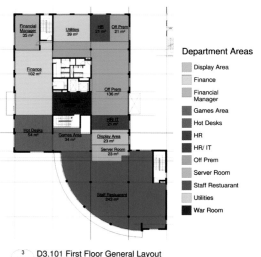

③ D3.101 First Floor General Layout
1:200

这个空间为大约 74 位职员设计，最多还可以再增加 15 人。科技是该设计的一个基本组成部分，因为红牛不仅要和他们的约翰内斯堡办事处互通资讯，更要与世界互通资讯。建筑师们把空间设计成工厂的样子，没有天花板，所有管线都裸露在外，同时也考虑到办公室的隔音需求。整个空间的建筑面积是按照鼓励员工在两层楼间自由活动的原则开展设计的，消除了办公空间中的层级感。

厨房部分是为员工享用午餐而设计，同时也可以用做娱乐活动室，体现真正的红牛精神。

在自然光下工作，并且使自然光充满整个办公空间是非常重要的。最终的设计成品是一个团队协作与隐私性共存的空间，并且噪音也得到了控制。现代与充满活力的设计展示了红牛品牌，并贯穿在整个空间中。

Capco Bold Rocket

Capco Bold Rocket 办公室

DESIGN COMPANY
D+DS Architecture Office
CLIENT
Capco Bold Rocket
LOCATION
London, UK
AREA
2,300 m^2
PHOTOGRAPHER
Tom Fallon Photography, Jeroen De Schrijver

The Capco Bold Rocket office in Shoreditch, London is located in a 150-year-old 5-storey steel and brick loft building.

Warm materials with high technology inserts, inspiring installations and an open plan create a stimulating and collaborative vibe.

The architectural and cultural context of the neighborhood inspire the design concept: open loft space, industrial architecture with character, solid materials, and visible joinery details. The custom furniture and walls use raw materials and expose the method of construction. Solid ash wood, timber pine either left natural or painted and black powder coated steel are the predominant materials. Walls reveal the method of construction, turning the typical construction inside out; wood studs are visible.

The building has a challenging shape, dividing the floor plan in 4 fairly separated corners with a central area in the middle. All floors have an open plan, in keeping with the loft character of the building.

The ground floor is the most public place. A series of volumes define different areas. The cafe is situated near the entry, creating a welcoming and relaxed feeling. A freestanding bleacher divides the entry from the central lounge. Facing the bleacher, another volume houses the main conference room – its interior painted in ochre to create a happy feeling even when the London weather is dreary outside – and it's exterior surface functions as a projection wall, turning the lounge & bleacher in a space where town hall presentations can be held.

Adjacent to the ochre room, a large orange rocket, which is a 3D print, is suspended in the stair void. The stair leads to the lower floor where the bold rocketeers work. Flanking the rocket, the wall of the fire staircase is transformed into a living wall, its plant patterns echoing billowing rocket smoke.

本项目办公室位于伦敦肖尔迪奇区一座拥有 150 年历史的五层钢砖结构厂房建筑中。

高科技的温和材料，令人振奋的内部装置和开放的设计营造了一种充满刺激感和合作感的氛围。

周围大楼的建筑以及文化背景启发了其设计概念：开放式的开敞空间，富有个性的工业建筑，牢固的材料以及可见的细木工艺。定制家具和墙面采用原生材料，将施工方式也展现出来。白蜡木、自然产或人为加工过的松木以及黑色粉末涂层钢板都是主要材料。墙面透露着这栋建筑的施工方式，平常建筑中的内部结构被暴露在外，使得木柱清晰可见。

大楼本身的形状也十分具有挑战性，它将空间分成四个独立的角落，中间是中心区。为了保持建筑的开敞特性，所有楼层都是开放式的设计。

第一层是公共区域，许多体块将其分为不同区域。咖啡厅就在入口，给人一种热情又轻松的感觉。一个立式的看台将入口和中心休息区分开。面向看台还有另外一个体块包括了主会议室 —— 它的内部喷上赭色，在伦敦这样潮湿的天气里，创造出愉悦的工作氛围；它的外部表面就好像是一面反射墙，将休息区和看台变成可以做市政厅展示的空间。

在赭色房间的旁边，有一个用立体打印技术制成的大型橘色火箭被悬挂在楼梯外。楼梯直接通往下层，那是大胆的火箭专家工作的地方。火箭旁边，靠近消防楼梯的墙面也被改造成生活化的墙面，上面的植物图案效仿了火箭翻滚的烟雾。

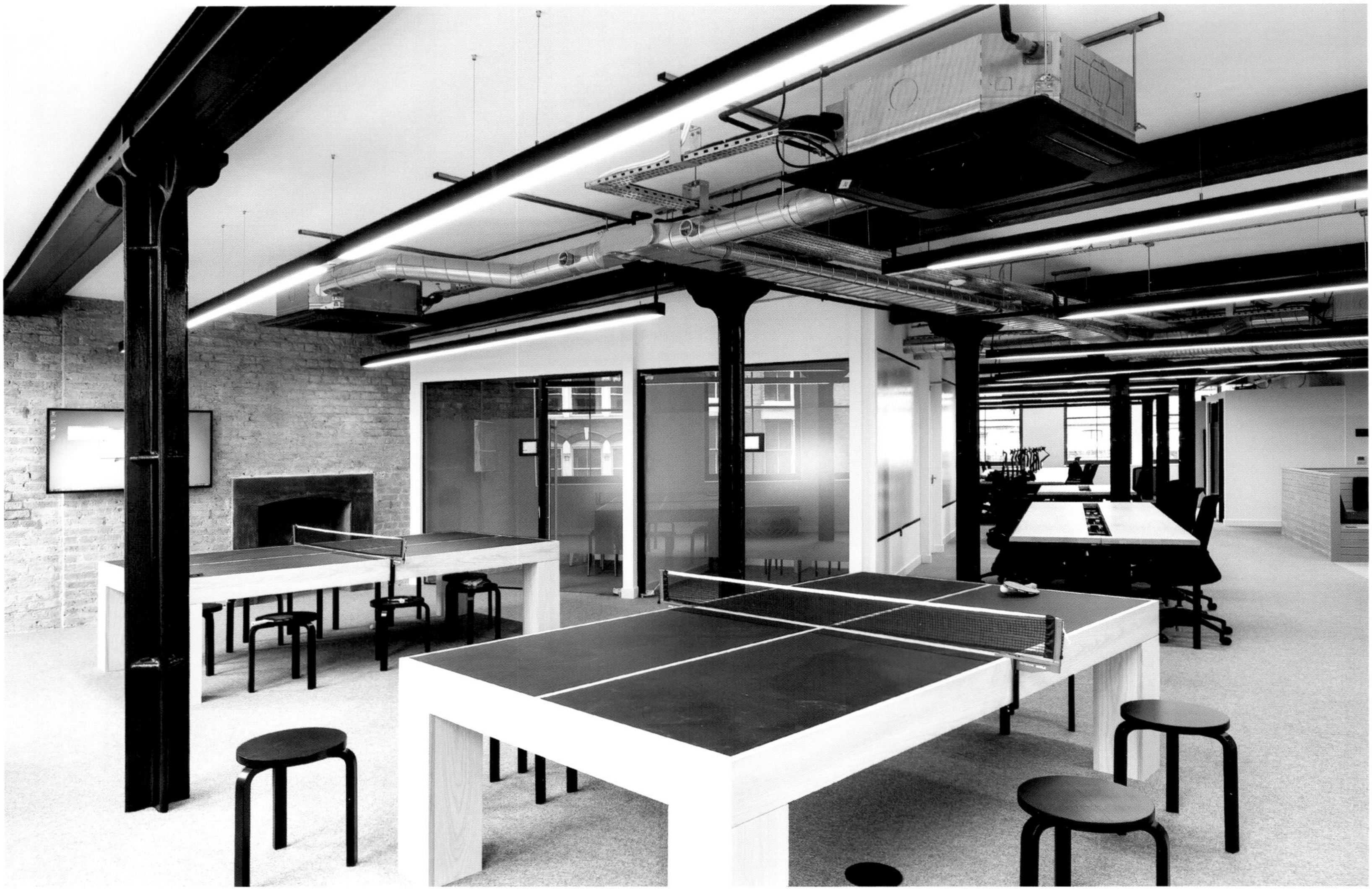

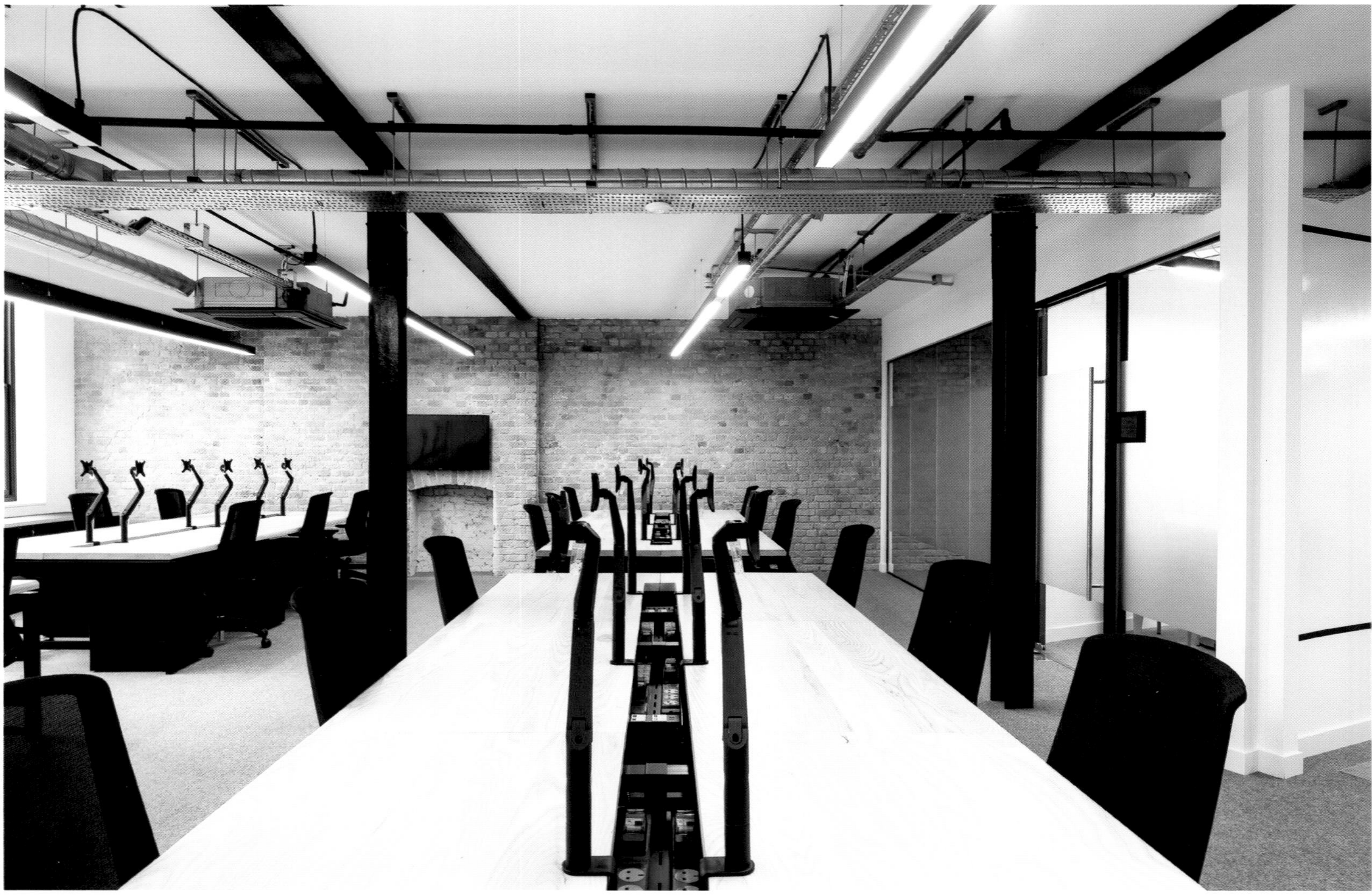

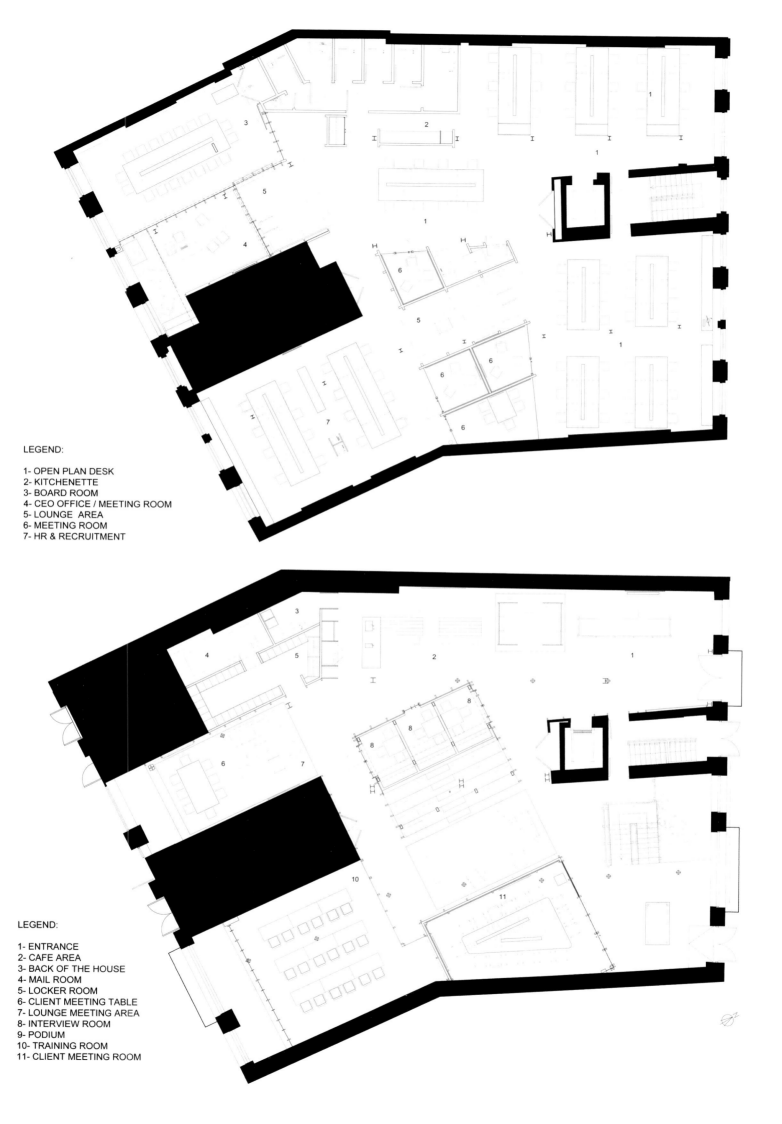

LEGEND:

1- OPEN PLAN DESK
2- KITCHENETTE
3- BOARD ROOM
4- CEO OFFICE / MEETING ROOM
5- LOUNGE AREA
6- MEETING ROOM
7- HR & RECRUITMENT

LEGEND:

1- ENTRANCE
2- CAFE AREA
3- BACK OF THE HOUSE
4- MAIL ROOM
5- LOCKER ROOM
6- CLIENT MEETING TABLE
7- LOUNGE MEETING AREA
8- INTERVIEW ROOM
9- PODIUM
10- TRAINING ROOM
11- CLIENT MEETING ROOM

Heavybit

Heavybit 办公空间

DESIGNER
Lisa Iwamoto, Craig Scott
DESIGN COMPANY
IwamotoScott Architecture
PROJECT TEAM
Sean Canty, Chretian Macutay
ASSISTANT
Ryan Beliakof, Anne Schneider,
Cooper Jones, Kelvin Huang

HEXCELL FABRIC CEILING INSTALLATION LEAD
Juliana Raimondi
LOCATION
San Francisco, California, U.S.A.
AREA
1,486.49 m²
PHOTOGRAPHER
Bruce Damonte

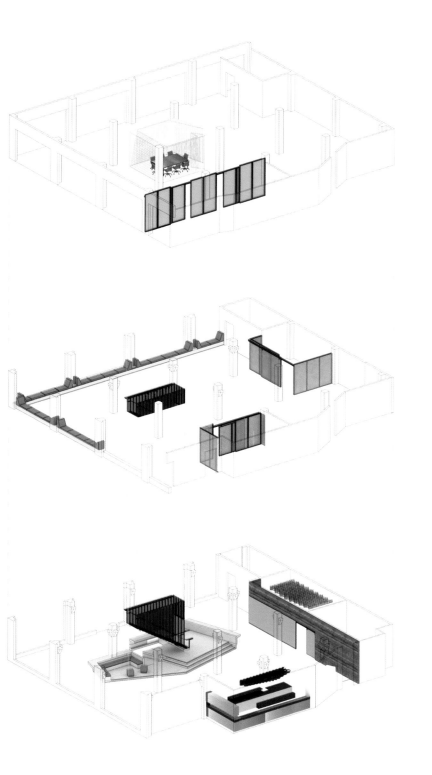

Heavybit Industries is a new workspace designed for early stage companies making cloud developer products. "HexCell Fabric Ceiling" is one of the three installations within the larger project were commissioned on a design-build basis, along with HexCell Steel Light and Rope Room. Both "HexCell" installations use a hexagonal plan pattern that recalls the Heavybit logo in different ways.

HexCell Fabric Ceiling is a ceiling light diffuser in the first floor conference room.

HexCell Fabric is a lightweight tensile ceiling structure. Made of ordinary and inexpensive non-woven mesh, the ceiling was designed using a physics modeler so that the flexible fabric is pulled taut equally in all directions, creating a geometrically precise but diffuse light effect in the room. In each case, as with the stair and platform, these installations attempt to defy the predictable qualities of the ordinary materials of which they are made.

Heavybit 产业是一种新型社交式办公空间，专为从事云开发产品制造的新公司而设计。"HexCell 结构天花板""HexCell 钢制灯"和"绳索屋"是其三大主要装置。其中，项目工程量较大的"HexCell 结构天花板"外包给了施工承包商。所有 HexCell 的装置设计都运用到了六边形的图案，旨在用不同的方式加深观众对 Heavybit 商标图形的记忆。

Hexcell 结构天花板是指位于一层会议室的天花板灯光发散装置。

Hexcell 结构是一种轻量级的拉伸天花板结构，这个天花板由普通廉价的无纺布网制成，设计时运用了一个物理模型，以便使灵活的织布可向四面八方平等拉紧，创造出一种从几何意义上布光精确却具有散光效果的天花板。在任何一种情况下，这些装置的设计，连同楼梯与平台的搭配，都在试图挑战设计原材料本身可预测的品质效果。

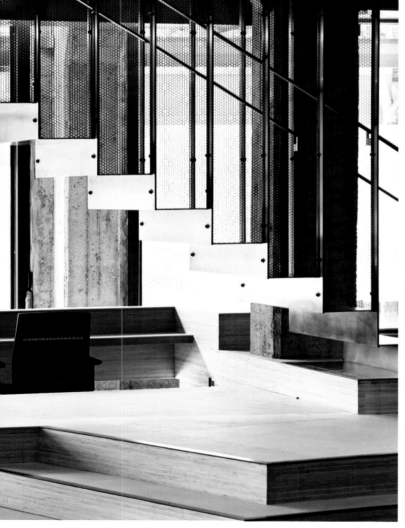

Office and Showroom of Architectural Workshop Sergey Makhno

谢尔盖·马克诺工作室办公室及陈列室

DESIGNER
Sergey Makhno, Ilya Tovstonog
DESIGN COMPANY
Sergey Makhno Architect
SCULPTOR
Nazar Bilyk, Dmitriy Grek, Sergey Redko, Yuriy Musatov
AREA
200 m²
PHOTOGRAPHER
Andrey Avdeenko

Minimalism with loft elements and warm notes of Ukrainian art create a special atmosphere in the workshop that, like a mirror, reflects the owner's inner being. The key concept of the architect's office project is to implement the elements worked out by the designers to meet their customers' needs.

The studio with a total area of 200 m² combines a showroom and a home-and-work place. In design Makhno used his favourite materials such as concrete, stone, glass, copper, bronze, various species of wood, high quality "TOTO" sanitary ware & fitting and "Miele" kitchen furniture.

The works of talented Ukrainian sculptors and artists create a special mood in the minimalist workshop. The main lighting ideas are conveyed based on Makhno's sketches. A bright spot on the stern and cold concrete wall is a lively green hedge-Epipremnum, the tropical plant from Holland. The flooring is considered to be ecologically pure and water-proof. Due to its smooth non-porous structure, it is void of cracks, joints and very resistant to bacteria.

Sergey Makhno cherishes the culture of Asia with rising sun, and honours a tea ceremony. For this reason, he designed a tea room in his studio where together with his friends and visitors the architect could enjoy a fragrant drink.

An important element in the workshop is a rich collection of Ukrainian zoomorphic ceramics. Makhno managed to pick out unique pottery pieces from different corners of Ukraine.

Lighting, room temperature and sound are controlled with the help of smart technology system by using a corporate mobile phone. Entering through the 3 m high door you will find a spacious business venue for conducting important events and meetings. It is also decorated with the author's chandeliers and wooden panels. The room is furnished with Sergey Makhno glass table and Kristalia "Elephant chairs". The collection of Ukrainian ceramics decorates the library which counts more than 1,000 books on design and architecture brought from all over the world.

KEY SECTION A-A

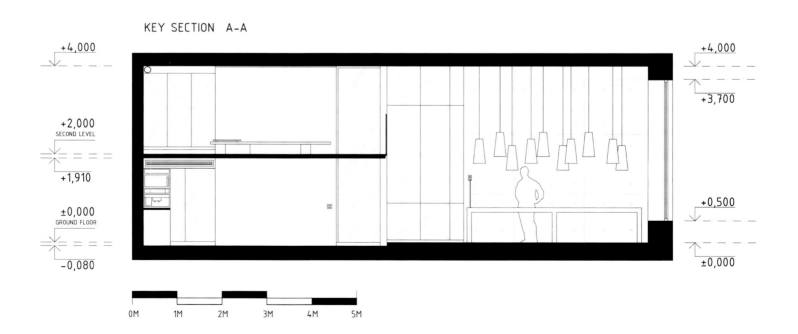

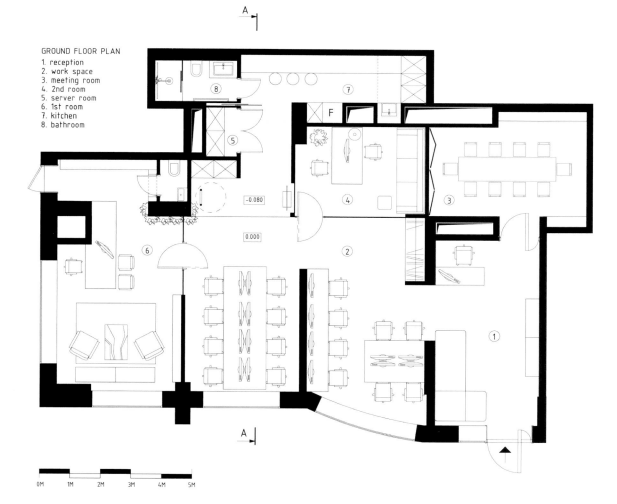

GROUND FLOOR PLAN
1. reception
2. work space
3. meeting room
4. 2nd room
5. server room
6. 1st room
7. kitchen
8. bathroom

利用 loft 元素和温馨的乌克兰艺术，设计师以极简主义风格在这间工作室中营造出了特别的气氛。这种情调如同一面镜子，反映着主人的内心世界。建筑师办公室项目的关键理念就在于如何实施设计师的方案来满足公司客户的需求。

工作室总面积为 200 m²，分为两个区域：陈列室和办公休息区。设计中马克诺使用了自己最喜欢的建材，例如，混凝土、石料、玻璃、青铜、各种木材以及高品质的东陶卫浴和美诺厨具。

出自乌克兰天才雕刻家和艺术家之手的艺术品为这间极简风格的工作室营造出了特别的气氛。主要照明设计的想法在马克诺的草图中得以传达。冷硬的混凝土墙上，有着生机勃勃的绿色树篱作为设计亮点。组成树篱的植物是来自荷兰的热带植物绿萝。地板选用了防水的纯生态材料，这种材料光滑无孔，不仅无缝防裂，还卫生抗菌。

谢尔盖·马克诺钟情于亚洲的"日升"文化和茶道艺术，因此他在办公室中特意设计了一个茶室，以供他和他的朋友及宾客们享受香茗。

设计中还有一个重要元素就是大量乌克兰动物陶器。马克诺从乌克兰的各个角落收集来了这些独特的物件。

由于采用了智能技术系统，使用手机便能够控制工作室内的照明设备、室内温度和音响。从 3 m 高的大门进入，你会看到一个专为重要活动和会议设计的广阔的商用空间。这里还装饰着设计师原创的支形吊灯和木质嵌板。房间里放置着谢尔盖·马克诺设计的玻璃办公桌和克里斯特拉的"大象椅"。乌克兰陶器收藏品布置在藏书室内，室内还有 1 000 多本来自全世界的设计和建筑类书籍。

SECOND LEVEL PLAN
+2,000 above zero
9. tea room

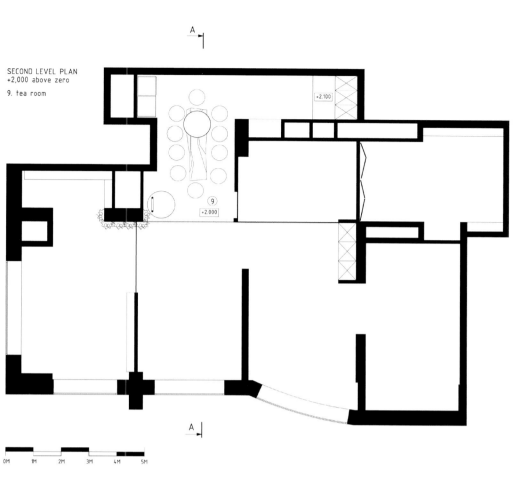

Red Bull Music Academy

红牛音乐学院

DESIGN COMPANY
INABA
DESIGN TEAM
Jeffrey Inaba, Ostap Rudakevych, Yoichiro Mizuno, Alan Kwan, Sean Connolly, Steven Tsa Shuning Zhao
EXECUTIVE ARCHITECT
Jill Leckner, Matthew Voss, Min Chen
LOCATION
New York, New York, U.S.A.
AREA
3,800 m^2
PHOTOGRAPHER
Greg Irikura, Naho Kubota

INABA was commissioned by Red Bull Music Academy to transform four floors of a vacant building in New York's Chelsea neighborhood into a model learning environment. INABA has used dramatic walls and lighting to create unique collaborative work areas in place of the traditional classroom.

Curving walls throughout the 3,800 m^2 forum give shape to the distinct spaces. On the ground level the walls extend far into the former warehouse providing passersby and users unrestricted views across the floor. They reveal a cross section of the types of activity taking place which include performances, private workshops, music production and broadcasting, aimed at making a statement that the space is different in its use from the shops, galleries, and cafés of the area. Below, on the cellar level, the arcing walls of the capsule-shaped lounge are interrupted only to establish long views from the recording studio located at the south end to the open-air patio at the north. On one of the upper floors, the similarly rounded walls enclose eight collaborative music studio pods. Each has large windows facing to open workrooms and the city skyline.

In an interior that is used at all hours of the day, the lighting plays a key role in setting the architectural atmosphere. During the day the ground level receives generous amounts of natural light from high floor-to-ceiling windows, while at night it is illuminated by rows of warm-colored custom-fabricated neon fixtures. Diffused LED lighting illuminates the radio studio and a programmable LED system focuses light on rows of acrylic tubes above the bar. Curved FRG light diffusers and indirect lighting lend an intimate setting to the auditorium while the ceiling structure supports light riggings for more theatrical effects. In smaller ancillary spaces, colored neon and grazing fixtures are employed in combination with high saturation paint. The cellar lounge has a low ceiling embedded with hundreds of linear LED fixtures to create a distributed field of light.

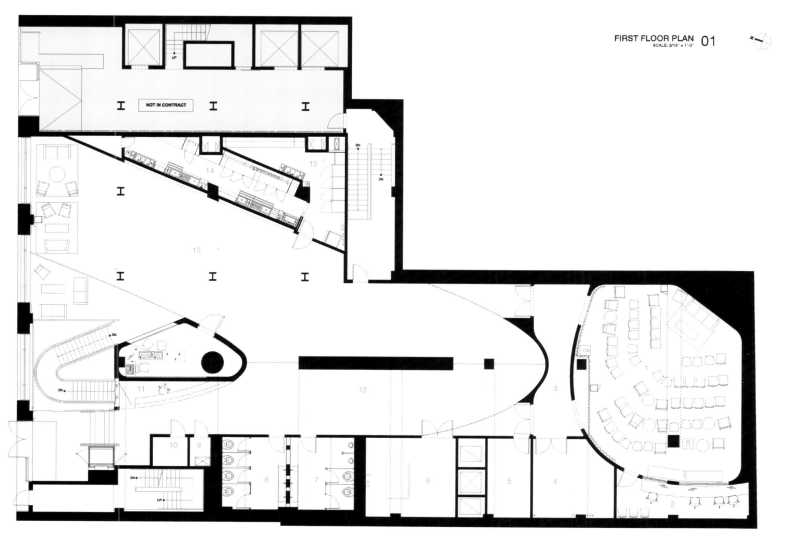

FIRST FLOOR PLAN 01
SCALE: 3/16" = 1'-0"

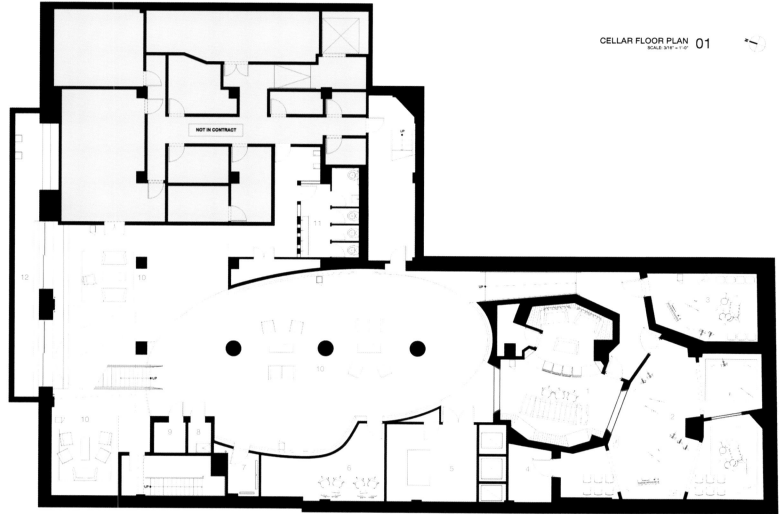

CELLAR FLOOR PLAN 01
SCALE: 3/16" = 1'-0"

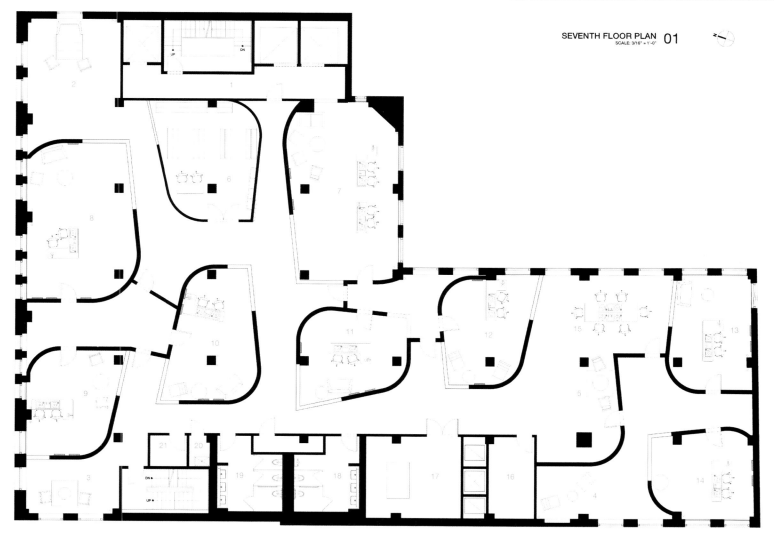

SEVENTH FLOOR PLAN 01
SCALE: 3/16" = 1'-0"

　　INABA 被红牛音乐学院委托将位于纽约切尔西区的四层空置建筑转变成一个模范学习环境。INABA 运用引人注目的墙面和灯光，营造出独特的合作工作环境，代替了传统的教室。

　　遍布在这个 3800 m² 的公共场所的弧形墙面给予了空间独特的形状。行走在一层，超大尺度的墙体一直延伸到以前的仓库，过往的行人和用户可以无障碍地欣赏整个楼层。开敞的空间可容纳多种功能，包括演出、私人工作室、音乐制作和广播，使该空间与周边区域的商店、画廊和咖啡厅在功能上有所区别。在地下一层，胶囊状休息室的弧形墙壁被切断，只是为了能够营造从南端的录音室直到北边的露天场地的广阔视野。上部楼层的一层中，相似的圆弧状墙体分隔出八个音乐制作室，并配以大型玻璃窗面向开敞的工作空间和美丽的城市天际线。

　　在这样一个全天候使用的建筑内，光成为制造空间氛围的主角。日间，一层通过高大的落地窗拥有充足的自然光照；晚间则是由几排暖光霓虹灯照明。散射的 LED 灯为录音工作室提供照明，可操控的 LED 光影系统将光对准吧台上的几排亚克力空心管。弯曲的 FRG 光扩散器和间接照明为礼堂提供了硬件设备，而其天花板结构支持照明设备营造出更多的戏剧效果。在较小的附属空间里，使用了彩色霓虹灯等设备与高饱和油漆相组合。地下休息室里较低的天花板主要以许多线形 LED 灯做装饰，形成一个光的世界。

SPARK Beijing Office
思邦北京办公室

DESIGN COMPANY
SPARK
PROJECT TEAM
Jan Felix Clostermann (Director), Christian Taeubert (Director), Li Lin (Architect)
LOCATION
Beijing, China
AREA
500 m²
PHOTOGRAPHER
He Shu

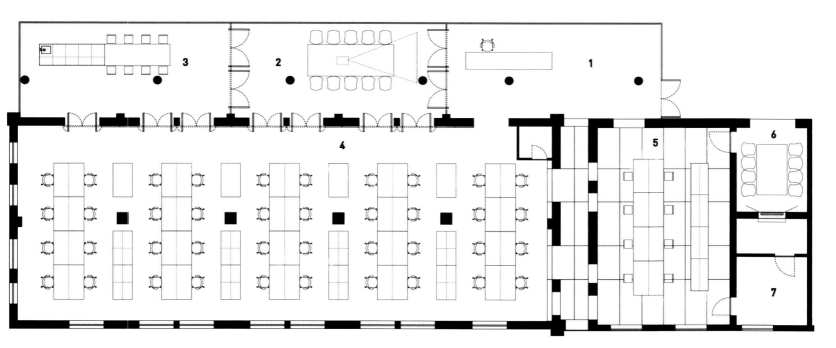

1. RECEPTION
2. MEETING ROOM
3. PANTRY
4. WORKING AREA
5. MODEL & MATERIAL LIBRARY
6. VIDEO MEETING ROOM
7. STORAGE & SERVER

The Liang Dian Design Center on Dongsishitiao is home to the new SPARK Beijing office. It occupies 500 m² area on the second floor of a three-storey office building built in the late 1980s.

An annex glass box has been added to second floor of the existing building, allowing for the exterior brick facade to be experienced as an interior space.

"We looked at the present configuration of the space and decided to keep it uninterrupted while inserting our office program," says SPARK Beijing's director Jan Felix Clostermann.

The space is loosely zoned into a main work area, kitchen and gathering space, a large meeting room area, a model making room, and smaller meeting room.

The window upstands of the existing brick facade facing towards the glass box are demolished, giving way to a much desired permeability of the space.

A continuous wall of metal-clad swivel doors along the side of the main work space allows for a playful and flexible configuration of "open and close".

When closed the metal panels catch subtle reflections of the surrounding office environment, turning it into a 20-meter-long "wall of imagination", inviting project teams to pin-up and discuss their work on its surface.

The model-making room, dubbed as "the ply room" is accessible via multiple routes, including through an extreme vertical space that used to be a stairwell core.

The flux of circulations between the reception area, the meeting room, and the kitchen can be adjusted via floor-to-ceiling swivel doors whenever required.

"The architecture profession fuels on idea exchange, so we sought to provide open collaborative spaces that are fun, where people want to be," says SPARK Beijing's director Christian Taeubert.

东四十条上的亮点设计中心是思邦北京新办公室的所在地。办公室占地面积为 500 m², 位于 20 世纪 80 年代末期建成的三层办公楼的第二层。

现有建筑楼的第二层增建了一个玻璃房，以便把外部砖砌外立面作为内部空间使用。

"在看过现有空间布局后，我们决定在进行办公室设计时，继续保留其原有的空间连贯性"，思邦北京办公室的主管简·费利克斯·克洛斯特曼说道。

这块空间会大致划分为主要工作区、厨房及聚会空间、大型会议室、模型制作室及小型会议室。

面对玻璃房现有砖墙外立面上的窗户的竖向构件被拆除，使空间的渗透性得到极大提高。

在主要工作室的旁边是几扇外部由金属面板包裹的连续旋转门，其构成的墙体使门的"一开一合"变得灵活而有趣。

关闭金属面板门后，可在门上发现周围办公环境的细微映像，使这几扇门变成一堵 20 m 长的"幻想墙"，邀请项目工作组的成员们聚在一起讨论工作。

模型制作室，也就是所谓的"板层室"，可以通过多条路径进入，其中包括一个曾用作楼梯井的极度垂直空间。

进出接待室、会议室以及厨房之间的人流量可以按需通过上下旋转门进行调整。

"建筑业需要交换不同的想法，因此我们试图提供一些有趣且让大家满意的开放性协作空间"，思邦北京办公室的主管克里斯琴·托伊帕说道。

AKQA

AKQA 办公室

DESIGNER
Craig Norman, Agustin Enriquez V
DESIGN COMPANY
GBD Architects Incorporated
LOCATION
Portland, Oregon, U.S.A.
AREA
1,077.67 m²
PHOTOGRAPHER
Peter Eckert

With offices around the world in major metropolises like Paris, Tokyo, London, Berlin, and Shanghai, AKQA was looking for an authentically Portland space to inspire their creativity. The Brewery Blocks in downtown Portland provided the perfect fit for their creative office space needs.

Prior to renovation, the lower level was a two-storey volume, featuring great bones and patina; the upper level was bland Class-A office space with corporate looking carpet tile, a lay-in acoustical ceiling and a dreary entry. GBD was tasked with removing the typical office quality of the space and allowing a century of heavy use to shine through.

GBD collaborated with a local builder for handcrafted, custom-built furniture for the reception desk, a large community table, and a beautiful conference table in the historic Brewmaster Suite. The design team prioritized locally-sourced and native materials, in order to reflect the unique qualities of Portland including new acoustical wood ceiling tiles, locally-harvested walnut, and refinished concrete flooring.

Completed in two phases – April 2013 and November 2014 – GBD designed a cohesive interior reflecting the creative culture of AKQA.

随着 AKQA 的办公室遍布于如巴黎、东京、伦敦、柏林和上海这样的中心城市，该公司也在物色真正位于波特兰市的一片天地，以此来激发他们的创造力。位于波特兰市中心的啤酒酿造厂街区恰好能满足他们创意办公空间的需求，为其提供了一个完美的落脚点。

翻修之前，较低层有两层楼的体积，以大型构架和铜绿色泽为特点；较高层就是普通的甲类办公室，配置有办公室专用方块毯、嵌入式吸声顶棚和常见门厅。GBD 建筑事务所担当起改变这个传统办公室空间的重任，让历经百年的空间变得光彩夺目。

GBD 建筑事务所制造出手工定制的接待台、大型会议桌和具有历史意义的酿酒师房间中的极美的会议桌。设计团队优先考虑使用当地的资源及材料，以凸显波特兰的独特气质，包括新的吸音木吊顶板、本地产的核桃木和整修过的混凝土地板。

其项目分为两个阶段——2013 年 4 月和 2014 年 11 月，GBD 建筑事务所设计的风格统一的室内环境反映了 AKQA 的创意文化。

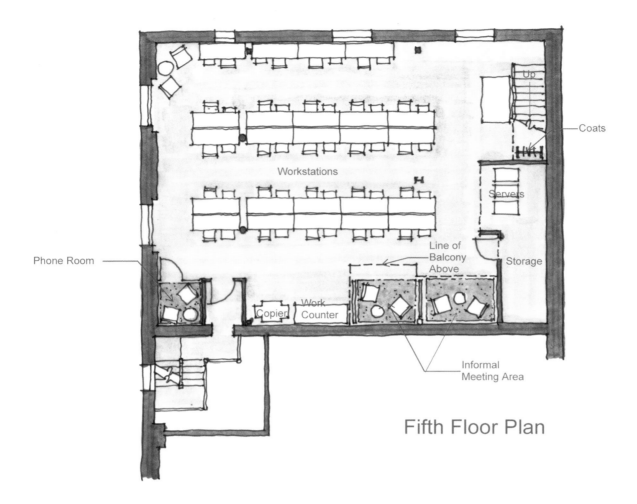

Fifth Floor Plan

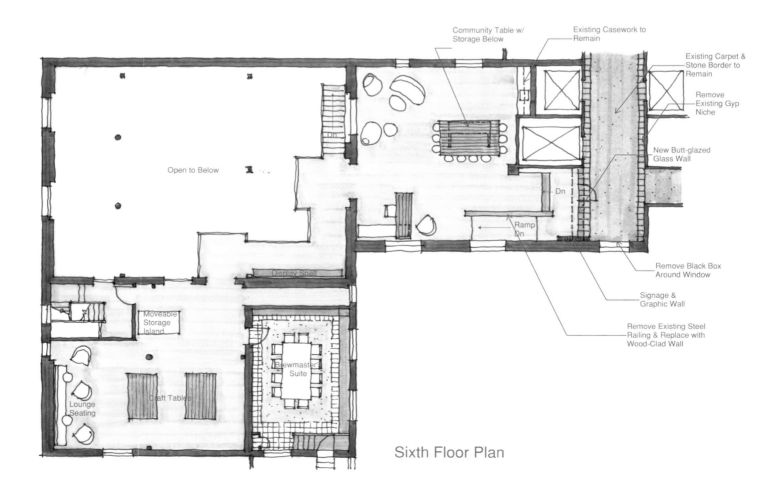

Sixth Floor Plan

Comcast Silicon Valley Innovation Center

DESIGN COMPANY
Design Blitz
LOCATION
Sunnyvale, California, U.S.A.
AREA
2,880 m²
PHOTOGRAPHER
Jasper Sanidad

康卡斯特硅谷创新中心

Located in Sunnyvale, Comcast's new Silicon Valley Innovation Center provides space for the rapidly growing company's entrepreneurial arm to build new products that change how people connect to entertainment, information, and their communities.

Needing an innovative and cool space for the Sunnyvale team, Comcast is partnered with Design Blitz to create a cutting-edge, collaborative new office.

The Comcast team is an especially bright and collaborative group, preferring to work in an open office environment which they refer to as their "innovation floor".

Cool, muted colors prevail over the space, with vibrant pockets of red throughout. Reflecting Comcast's origins as a cable provider, the design of the space takes cues from a two-dimensional electrical wiring diagram. Red paths run across the floor using in the pattern of wires and serve as way-finding guidelines. The architects further extrapolate the pattern into a three-dimensional form, creating red hooded structures which divide the space and provide alternate collaboration and meeting areas.

While the new Innovation Center is tech-centric, it is important for the project to incorporate green technology and be environmentally friendly. Materials with multiple uses were selected throughout, such as ones that had privacy, acoustic, and aesthetic functions. The project has received LEED Gold certification.

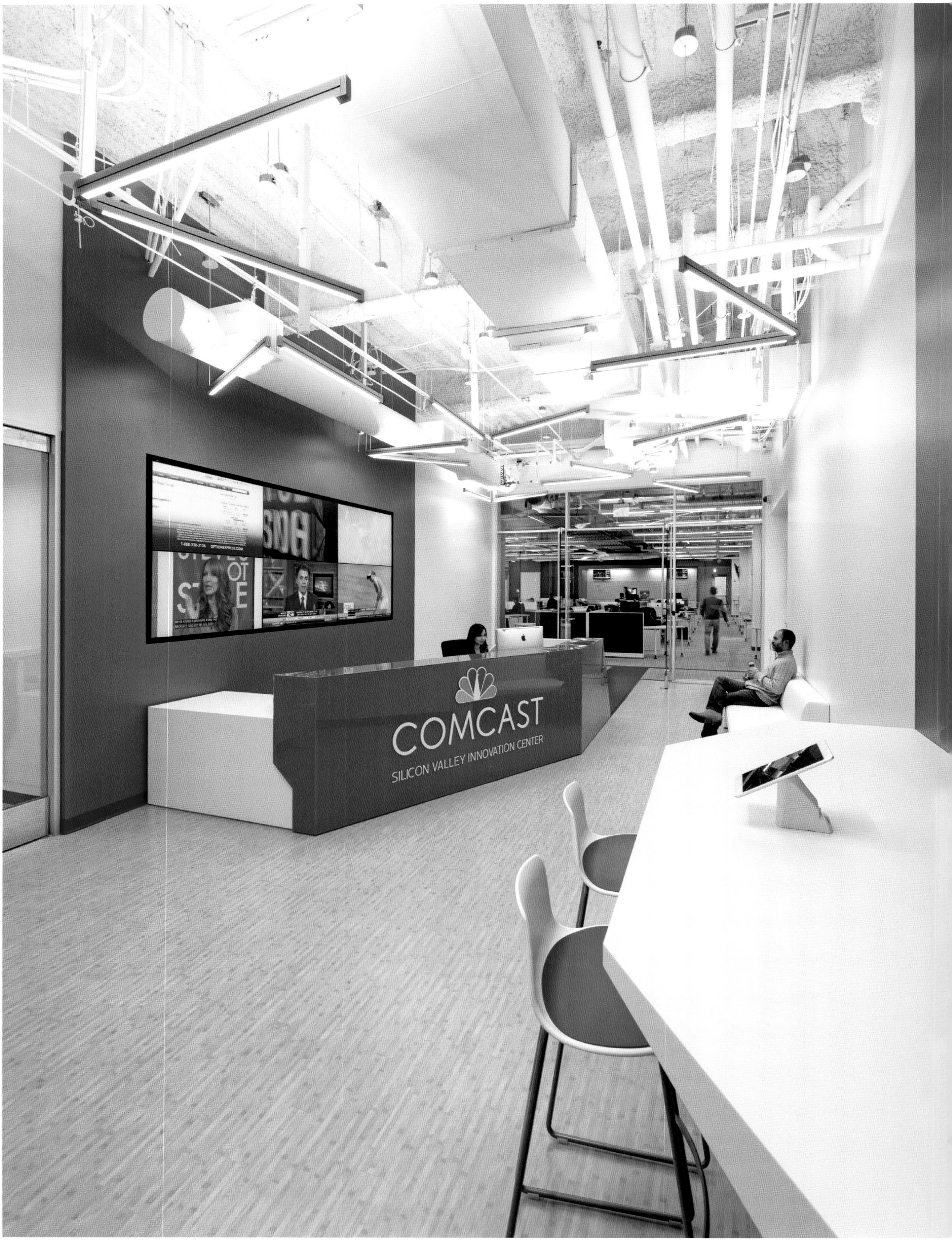
COMCAST
SILICON VALLEY INNOVATION CENTER

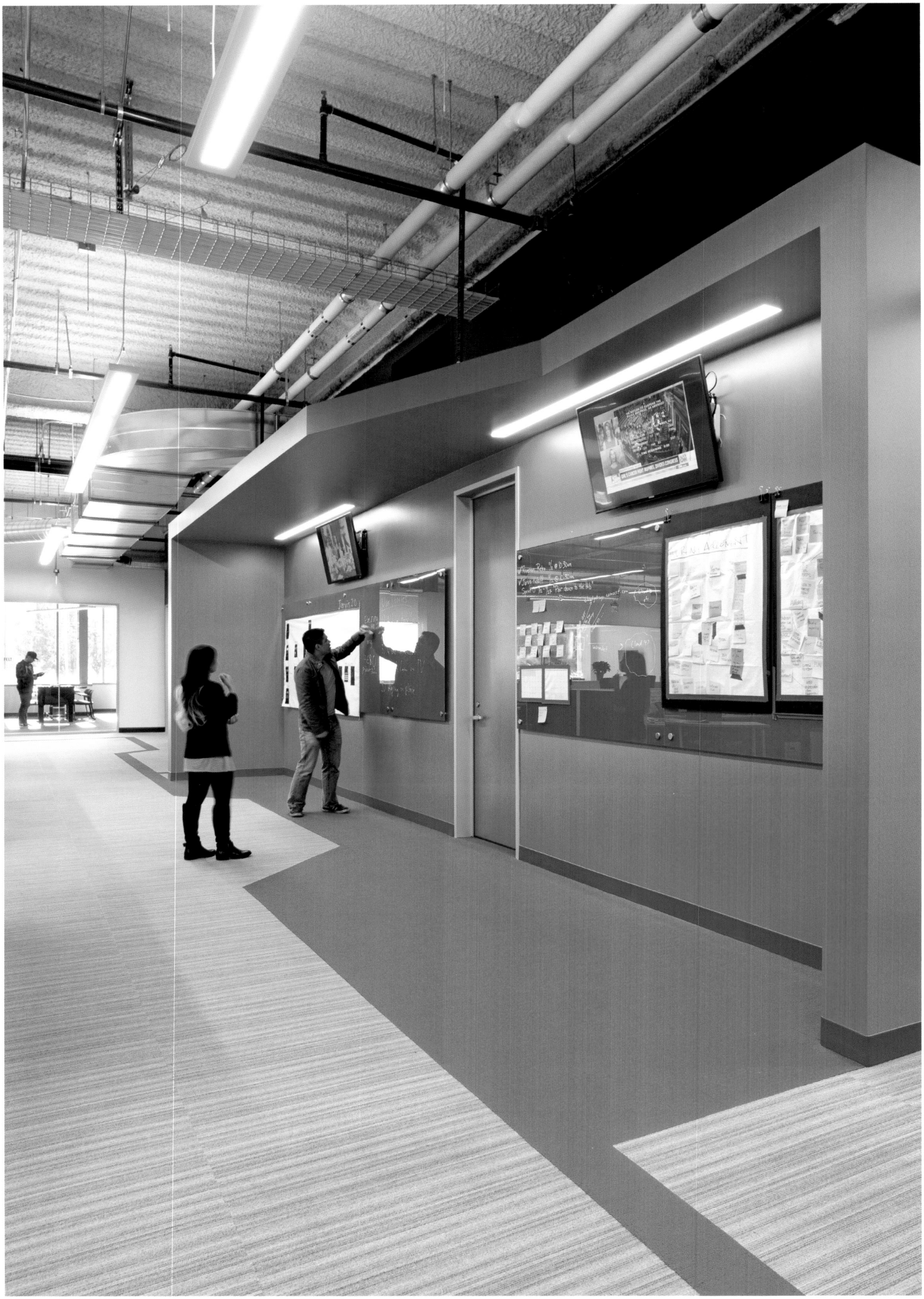

康卡斯特的新硅谷创新中心位于森尼韦尔市，它为公司新产品的开发提供了理想场所。这些产品可以改变人们娱乐和获取信息的方式，以及他们的社交圈。

森尼韦尔团队急需一处既有创意又时尚的地方，于是，康卡斯特和布利茨设计事务所合作打造了一间与时代前沿紧密结合的新型办公室。

康卡斯特团队是一个特别充满活力、具有协作精神的集体，喜欢在开放式的办公环境中工作，他们称之为"创新楼层"。

清爽、柔和的色彩充斥着空间，鲜亮醒目的红色映衬其间。为了反映出康卡斯特公司作为电缆供应商的起源，空间设计按照一幅二维平面电气线路图进行设计。红色路线横跨于带有线路图案的地板上，并作为路标。建筑师进一步将平面模式处理成了立体模式，打造出红色建筑体，并以其划分了空间，提供了可交替使用的合作区域和会议区域。

新的创新中心虽然是以科技为中心，但是使项目结合绿色技术以及环保也是很重要的。多用途材料都经过了精心挑选，例如，有些材料就兼具私密性、良好的建筑声学和美学特性。该项目获美国绿色建筑评估体系 LEED 金牌认证。

Corporate Office in Milan

米兰分公司

DESIGNER
Maurizio Lai
DESIGN COMPANY
Lai Studio
PROJECT TEAM
Maurizio Lai (Senior Architect), Giuseppe Tallarita (Architect), Michele Capra (Architect), Marco de Santi (Architect), Elena Mazzoleni (Material Selection & Styling), Beatriz Jam de Leon (Junior Designer), Nella Figueroa (Junior Designer)
CONSULTANT
Riccardo Schironi (Engineering)
CONTRACTOR
Magistri Costruzioni Srl, Ciro Monguzzi, LCS (Laboratorio Costruzioni Scenografiche), Talmax Oltrevetro Climarredo
LOCATION
Milan, Italy
AREA
610 m² (Built)
PHOTOGRAPHER
Andrea Martiradonna

The recovery of a semi-industrial space leads to the conception of an environment by intersecting volumes, where floors and surfaces immerse in each other, looking for the light.

Initially a mechanical workshop, these premises without windows, used to get the light from the large windows at the rear, overlooking a small courtyard, as well as through skylights from the ceiling.

The project of restoration and transformation of the space in the representative office of a real estate development company grows around the need of maximization of natural light.

On the ground floor, the entrance is marked by a scenic 17 m long corridor, highlighted by recessed light strips that run along the ceiling and the floor, where a geometric pattern of false perspective deceives the eyes of the visitors.

Along the corridor, a large furnished niche serves as the main waiting area, closed at the bottom by a large sliding wall that works as a second entrance.

The Executive Offices overlook a glass catwalk that delimits the fragmented portion of the ground floor and ends in the Office of the Presidency, a great volume provided with its own Conference Room, where the vertical garden provides a lush green background through the large glass window.

The main staircase that winds its way from the catwalk looks like a great metal band on the outside: it accommodates pale color steps made of kerlite slabs and light holes along the way, which are reflected to infinity in the mirror covered parapets, creating a kaleidoscopic effect.

The staircase leads to the basement where the Architecture and Design Department of the Company is located, divided into a long strip of workstations and an area used for internal meetings and design reviews.

The secondary staircase, instead, connects the Administrative Offices positioned at the rear of the basement with the Management upstairs.

A third staircase leads to a further basement, where the archives and the technical facilities are.

Sezione C-C'

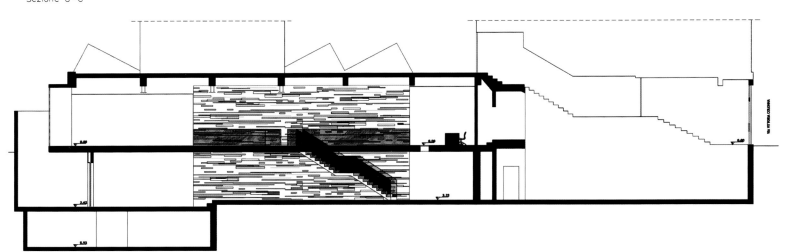

Sezione A-A'

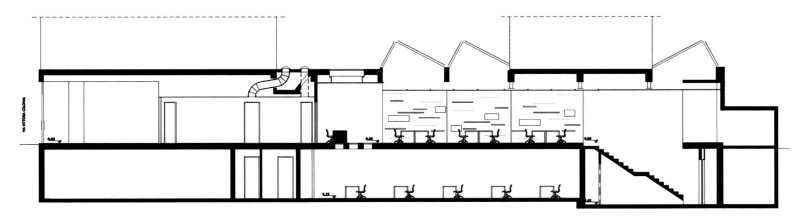

PIANTE, PROSPETTI E SEZIONE

PROSPETTIVA

Sezione B-B'

Sezione F-F'

Prospetto su via V. Colonna

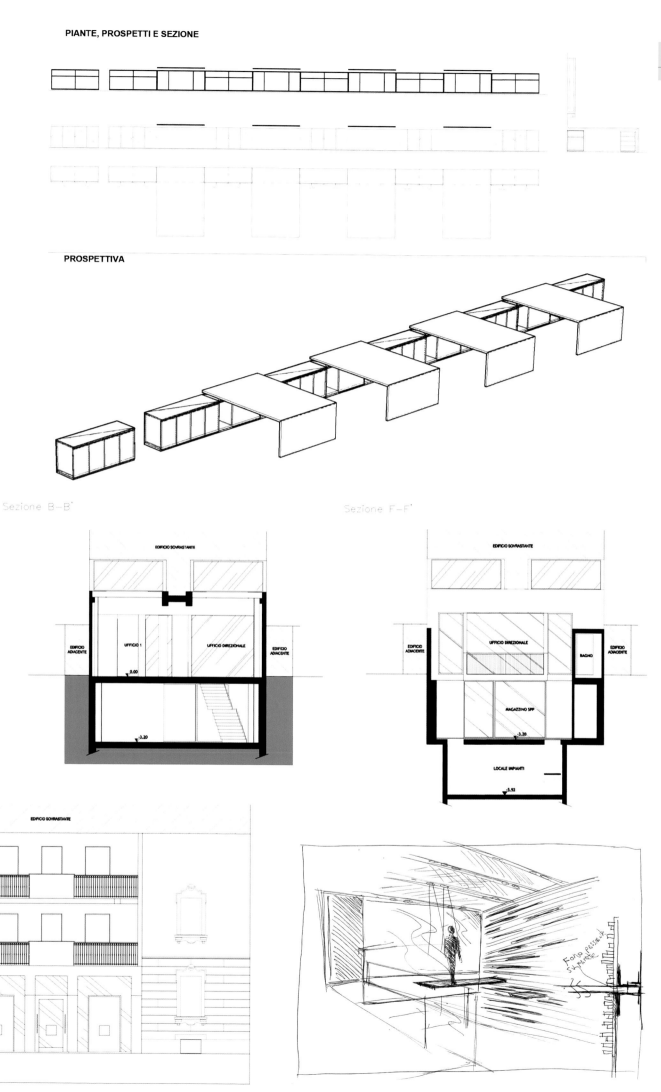

GROUND FLOOR

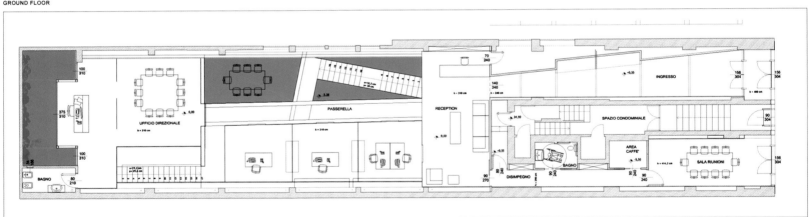

FIRST UNDERGROUND FLOOR

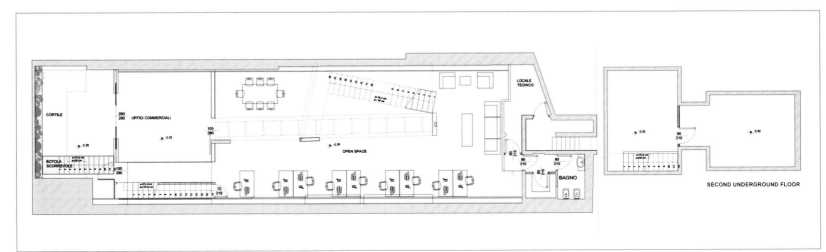

SECOND UNDERGROUND FLOOR

  半工业性空间的复原产生了交叉相错的建筑体量设计，地面和表面相互遮盖渗透，以寻找光源。

  起初作为机械工厂，这些建筑没有窗户，采光来源于后面面向远处小庭院的大型通风窗，也来自天花板上的天窗。

  因一家房地产开发公司办事处的场地重整项目，自然采光的需求变得更加迫切。

  一层入口处有一条长 17 m 的漂亮走廊，嵌入式灯条安装在天花板和地板上，其中的假透视几何图案让游客目不暇接。

  沿着走廊有一个充当主等候区的大型凹槽，其尽头是一面滑动墙板，用作第二入口。

  从行政办公室可俯瞰一座玻璃天桥，这座桥把一层的零散区域隔开。桥的尽头是总裁办公室，其拥有巨大的建筑体量，并配有会议室，透过一扇很大的玻璃窗，可以看到那里的垂直花园投下一片郁郁葱葱的绿色。

  主楼梯从天桥蜿蜒而下，看起来就像盘在外面的金属带。楼梯将由 Kerlite 陶瓷薄板制成的浅色台阶和延路的灯孔相搭配，灯光反射到护栏前的镜子里把景物拉到无限远，营造出一种万花筒似的效果。

  楼梯通向地下室，那里是公司建筑与设计部的所在地，被划分为一长排工作室和一处内部会议和设计评审区。

  而相反地，第二段楼梯把地下室后方的行政办公室与楼上的管理部门连接起来。

  第三段楼梯通向更远处的地下室，那里用于存放档案和技术设施。

Corus Quay

哥鲁氏码头办公室

DESIGN COMPANY
Quadrangle Architects
PROJECT TEAM
Brian Curtner, Rob Dyson, Caroline Robbie, Ted Shore, Panyada Wangpongpipat, Frances Hahn, Vera Gisarov, Jennifer Heimpel, Michelle Xuereb, Young Kun Yoon, Andrea Shearer, Rick Mugford, Julie Mroczkowski, Jennifer Lembke, Viviane Chan, Ana Francisca De La Mora, Kateryna Nebesna

LOCATION
Toronto, Ontario, Canada
AREA
44,500.55 m²
PHOTOGRAPHER
Ben Rahn (A Frame)

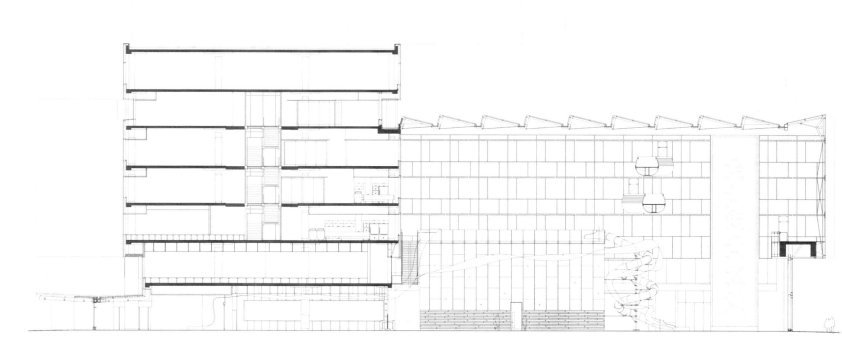

BUILDING SECTION

Corus commissioned Quadrangle Architects to help create a centralized office that would flaunt technology and sustainability, heighten engagement and creativity, and to simultaneously express a confident and unified corporate identity and the bold character of each of Corus' brands.

Within a new building commissioned by the City of Toronto (designed by Diamond Schmitt Architects) on the edge of Lake Ontario, Corus' office incorporates serious fun into the workplace. Openness and transparency abound in the space. Open floorplates, raw ceilings, exposed ducts and cabling give the offices a loft-like feeling. There are no closed offices. Staff work from desks divided by brightly-colored partitions in waves referencing the waterfront location. A glass roof over a central atrium and floor-to-ceiling glazing on all external walls maximize natural light. Boardrooms, telephone rooms, editing suites and banks of electronics are also "open", showcased behind glazed walls. Radio broadcasting booths are exposed to the adjacent Sugar Beach Park, connecting the broadcasters with their public and giving broad views to employees used to working in dark, enclosed spaces.

While the interior reflects the fast-pace of multi-media culture, Corus' distance from the bustle of downtown and its connections to the lake, the beach and the sky pervade the interior with tranquility. And the design is not just about image. It is designed for long-term flexibility and sustainability, targeting LEED® ND Gold.

Each floor has a lively kitchen, breakout spaces and there are generous outdoor terraces that invite staff to meet informally. The most significant of these gathering places is on the roof of the TV production studios within the building at the heart of the atrium. There employees can plug in around a Steelcase Mediascape lounge, sketch under oversized Luxo Lamps, play football and even release stress by taking a ride down the slide to the ground floor. The slide has become an important symbol for Corus – as a delightful and inspiring place in which to work.

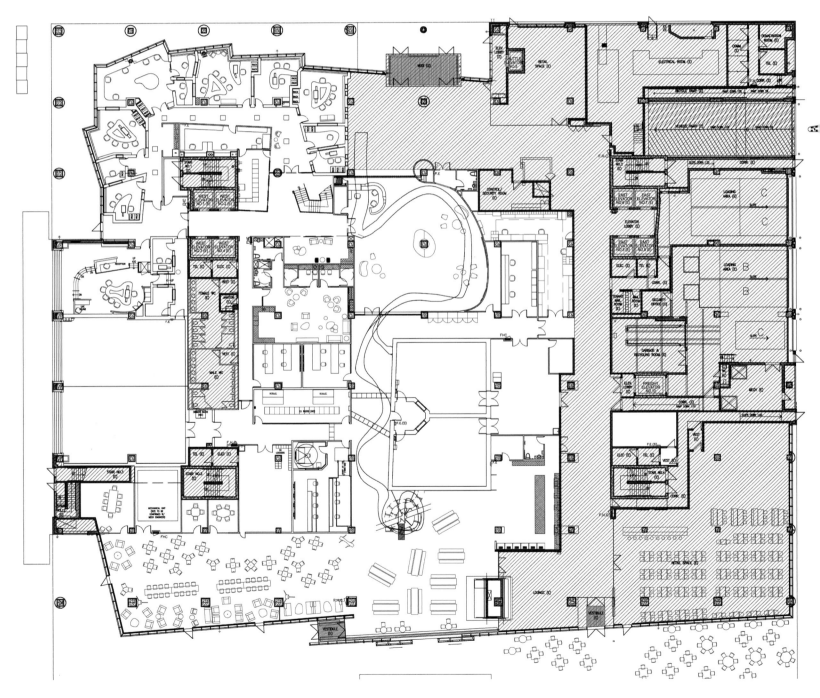

GROUND FLOOR + PUBLIC STUDIOS

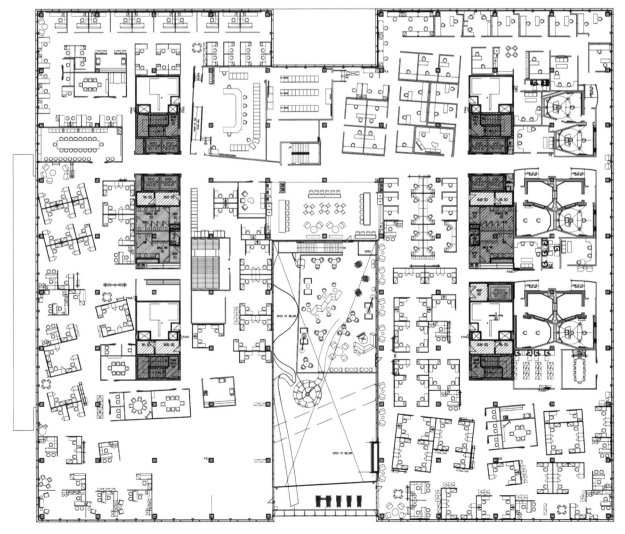

THIRD FLOOR
COMBINED TYPICAL OFFICE
AND TECHNICAL SPACES

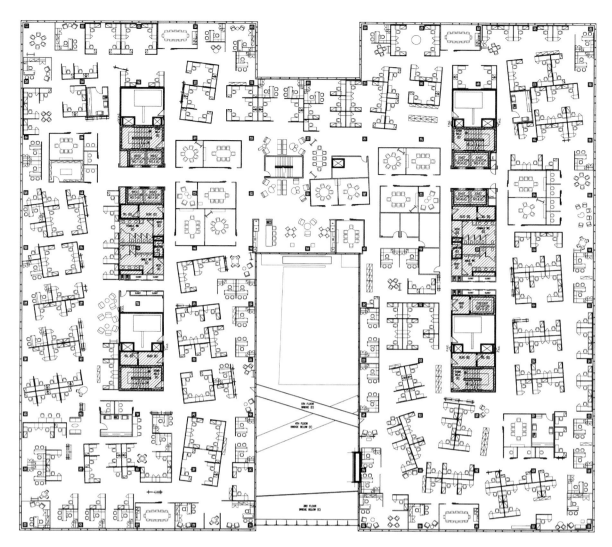

FIFTH FLOOR
TYPICAL OPEN OFFICE FLOOR
WITH COLLABORATION ZONES
AND GLAZED MEETING ROOMS

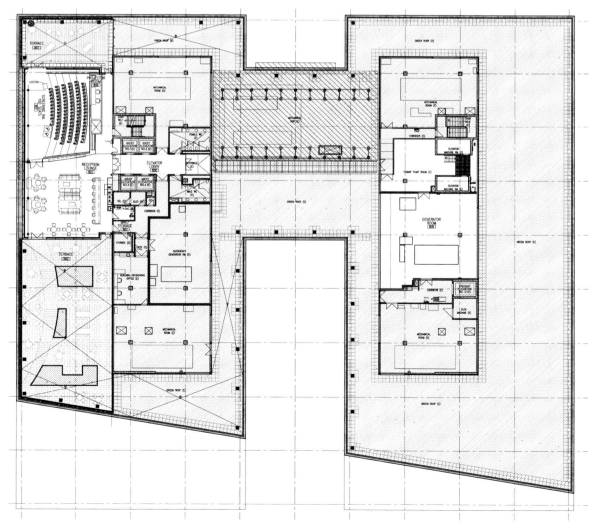

EIGHTH FLOOR
EVENT SPACE AND GREEN
ROOF

哥鲁氏公司委托 Quadrangle 建筑设计有限公司来创建一个集中性强的办公室，以此标榜技术实力和持续性，提高参与性和创造性，同时表现出其自信统一的企业形象以及哥鲁氏旗下各品牌的大胆特色。

由多伦多市政府委托建造，位于安大略湖湖畔的新大楼（由 Diamond Schmitt 事务所设计）内，哥鲁氏的办公室严肃而不失乐趣。整个空间具有十足的公开性和透明性。开阔宽敞的地板、原生态的天花板、裸露在外的管道和电缆赋予办公室一种高挑开敞的感觉。这里并没有紧闭的办公室。员工们在办公桌前工作，这些办公桌参照滨水地区的位置特色用颜色鲜亮的隔板呈波浪式隔开。中庭中间的玻璃屋顶和所有外墙的落地玻璃使自然采光得以最大化。会议室、电话室、编辑室和电子设备间也都是"开放"的，展示在玻璃墙后面。无线电广播室朝向相邻的糖果湖滨公园，将广播员与大众联系起来，给通常要在封闭黑暗场所工作的员工开阔的视野。

室内体现的是多媒体文化的快节奏感，然而，哥鲁氏远离喧嚣的市中心，贴近湖水、湖滨和蓝天使得其内部充满宁静的气息。而且其设计不仅关乎形象问题，还要为长期的灵活性和可持续性考虑，并以获得美国绿色建筑社区评估体系（LEED ND）认证为目标。

每一层都配备有一个温馨可爱的厨房、休闲空间和宽敞可以邀请职员来开非正式会议的户外阳台。这些空间最大的亮点是在电视制作室的屋顶——坐落在一栋建筑物内中庭的中心位置。员工可以在 Steelcase 办公家具公司媒体景观（Mediascape）休息室附近充电，在大型 Luxo 台灯下描绘图，玩玩桌式足球甚至能坐滑梯到一层释放压力。这座滑梯已经成为哥鲁氏的重要标志，使这里成为既愉快又令人振奋的工作地点。

De Burgemeester

De Burgemeester 办公大楼

DESIGN COMPANY
Studioninedots
LOCATION
Burgemeester Pabstlaan, Hoofddorp, The Netherlands
PHOTOGRAPHER
Peter Cuypers

From an abandoned and anonymous office building to a vibrant multi-tenant complex, that's De Burgemeester, a commercial property in Hoofddorp renovated by Studioninedots and opened on 5 November 2013. What's the secret of the transformation? A vertical lobby features an open staircase where people meet face to face and a space that brings people together both literally and figuratively.

We space – that's the name for this communal area at the heart of the building. It's a place that brings people together. Out of the concrete floors the architects carved a 14-meter-tall void that houses a giant staircase which cuts diagonal lines through the void as it makes its way upwards, linking the different floors to one another. Now people are on the move, making their way back and forth on the timber steps. Some of them linger for a chat, and there's space on the broad treads to sit for a moment. The sound of chatter and the aroma of coffee from the café below now fill the hall. Most of the office space has already been leased, bringing the building back to life once again.

What's more, the crisis offers opportunities. The pressure of cost-cutting measures creates scope for other values. More does not necessarily mean better, and that's something increasing numbers of people are coming to realize. Born out of the idea of cutbacks and facilitated by the internet, a flourishing culture of sharing has emerged. Thanks to Greenwheels, Peerby and Airbnb, people borrow and rent cars, tools and even homes from one another. People in more and more cities are setting up resident associations to make their neighborhoods more sustainable. And vacant sites are taken as "test sites" for new spatial developments such as urban farming. The sense that "everybody for himself" no longer works, and the feeling that we can really achieve something by joining forces, is gaining widespread support. And, just as important: we've come to realize that it's much more fun together.

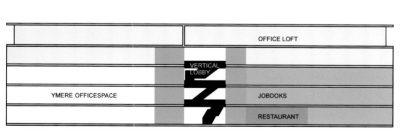

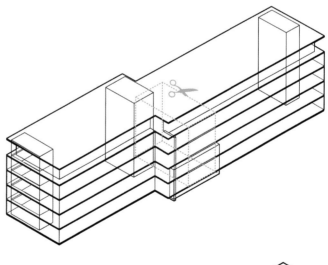

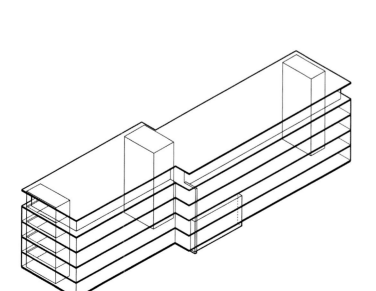

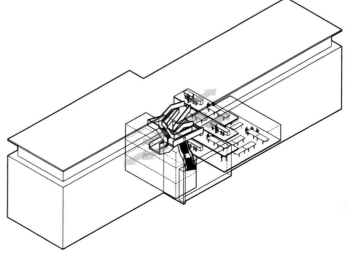

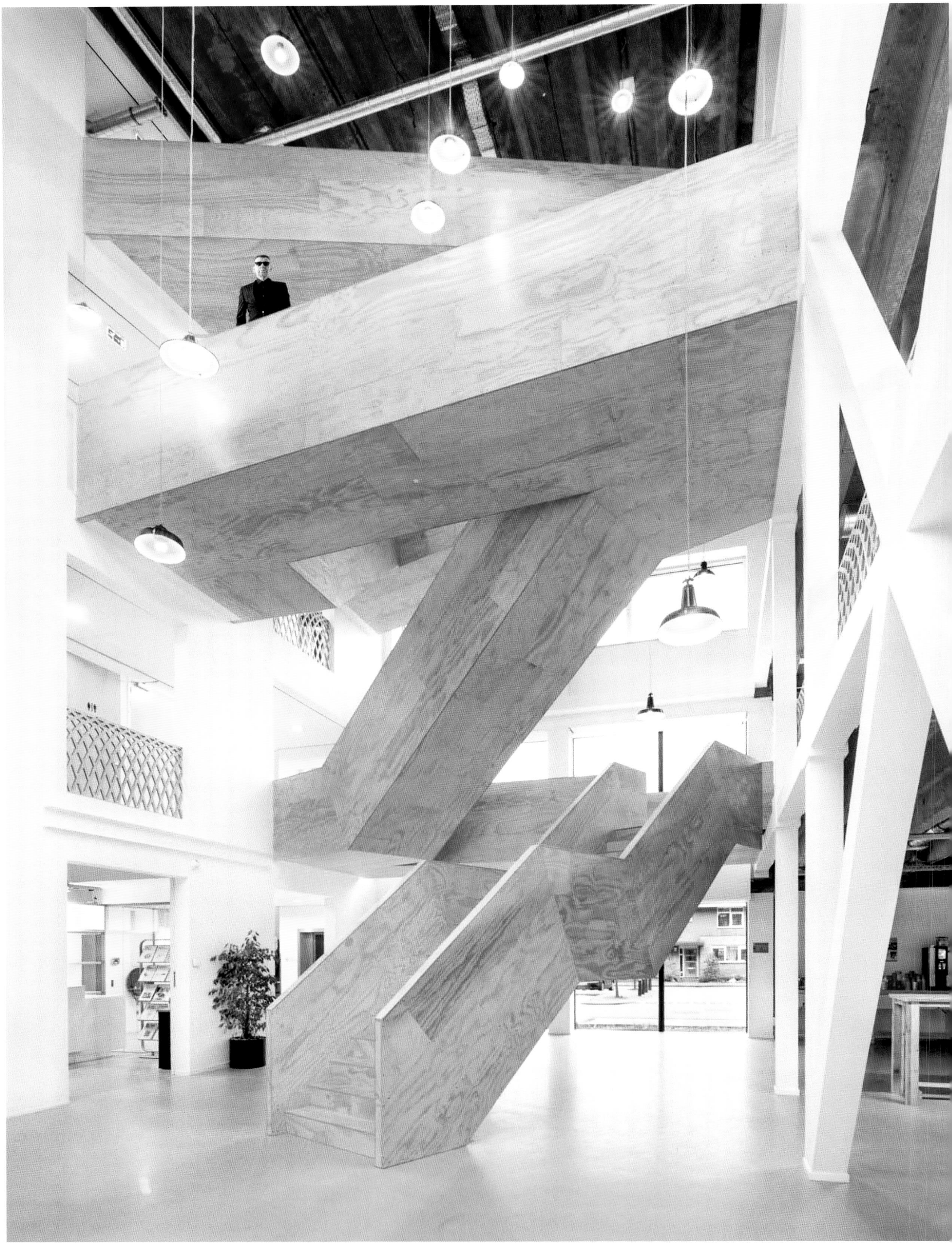

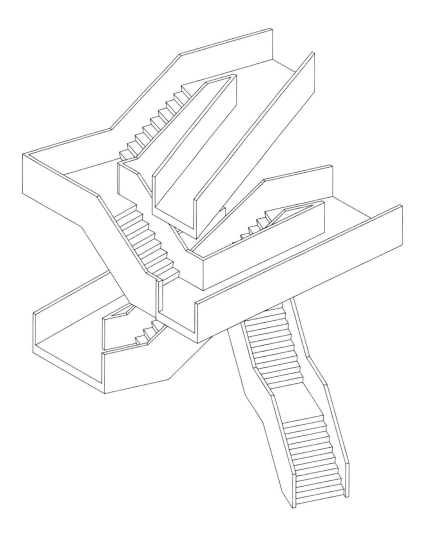

　　De Burgemeester 的前身是一座无名的废弃大楼，现在是一个朝气蓬勃的多租户办公大楼。De Burgemeester，这座位于荷兰霍夫多普的商业地产由 Studioninedots 设计团队翻修，2013 年 11 月 5 号开放使用。大楼改造的秘诀是什么？秘诀就是一个垂直的大厅并建有使人们可以面对面交流的敞开式的楼梯，以及一个把人们自由灵活地聚在一起的空间。

　　"我们的空间"，这是建筑设计师们给大楼中心公共区域起的名字。这个区域将人们聚集在一起。建筑设计师们在水泥地面上建了一个 14 m 高的空间来做巨大的楼梯，以对角线形式在空间穿越，将不同楼层连接起来。现在人们四处走动，通过木质楼梯到他们想去的楼层。一些人在楼梯上驻足交谈，宽阔的台阶也可以让人们坐下来休息片刻。人们的交谈声和楼下咖啡馆的咖啡香气弥漫在整个大厅中。现在大部分的办公室空间都已出租，整栋大楼再次充满生机。

　　除此之外，建筑设计的挑战也带来了机遇。削减成本的压力激发了其他的价值。越来越多的人们认识到更多并不意味着更好。源于节约观念，并由互联网推动，"分享文化"得以出现和盛行。感谢 Greenwheels、Peerby 和 Airbnb 公司的出现，人们可以租借汽车、工具，甚至还有房屋。现在越来越多的城市中的人们建立起居民协会使自己的社区更好地实现可持续发展。同时，为了新的空间发展，许多空置地盘用来做试验场地，例如，城市农场。人们也不再提倡"人人为己"的思想，团结一致达到目标这一观念得到广泛的支持。同样重要的是，人们也认识到只有大家聚在一起才能发掘更多的乐趣。

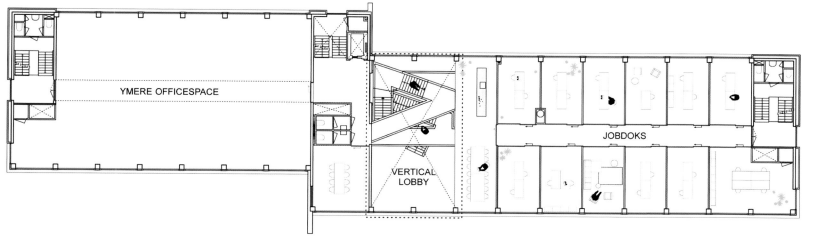

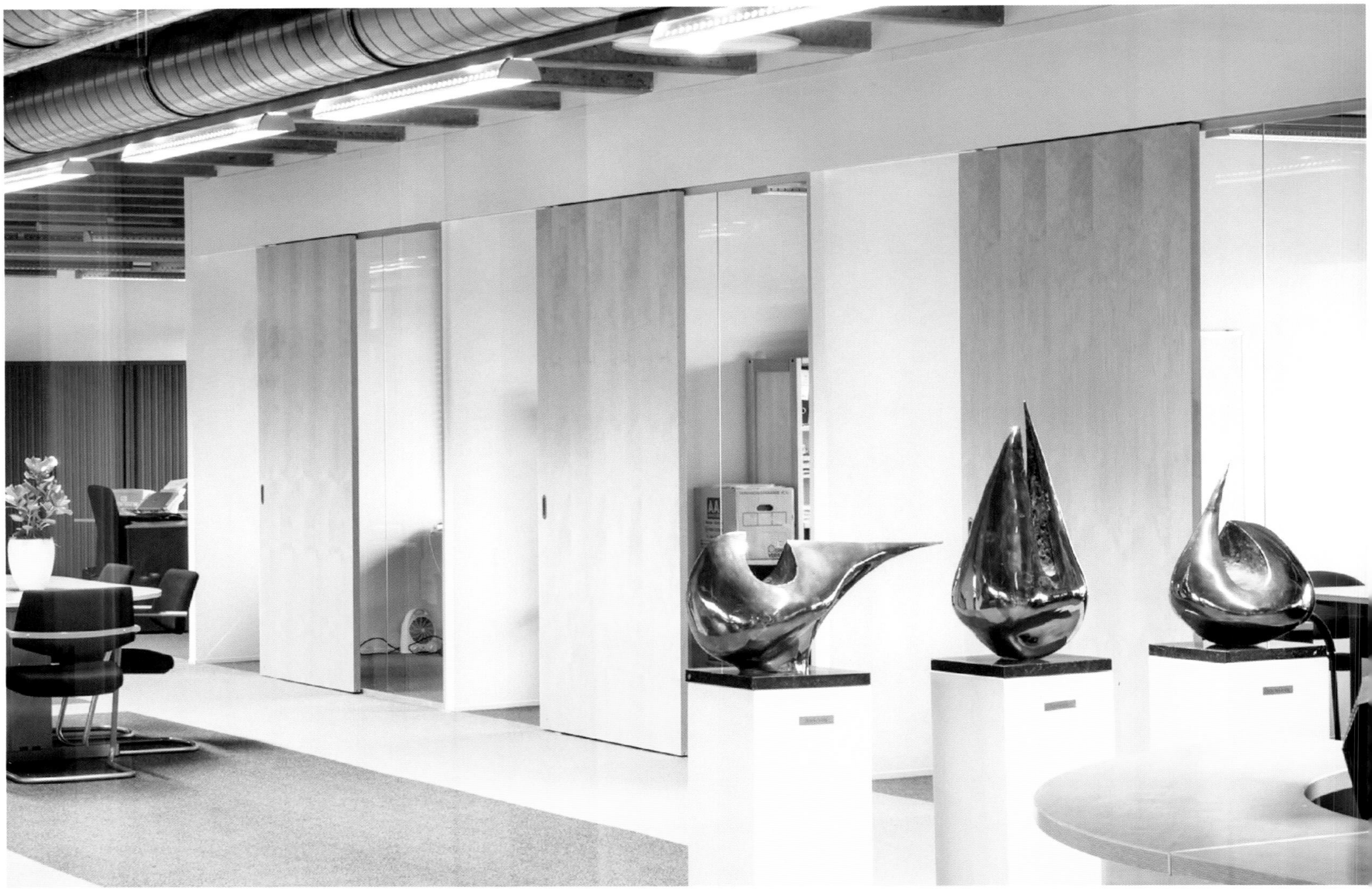

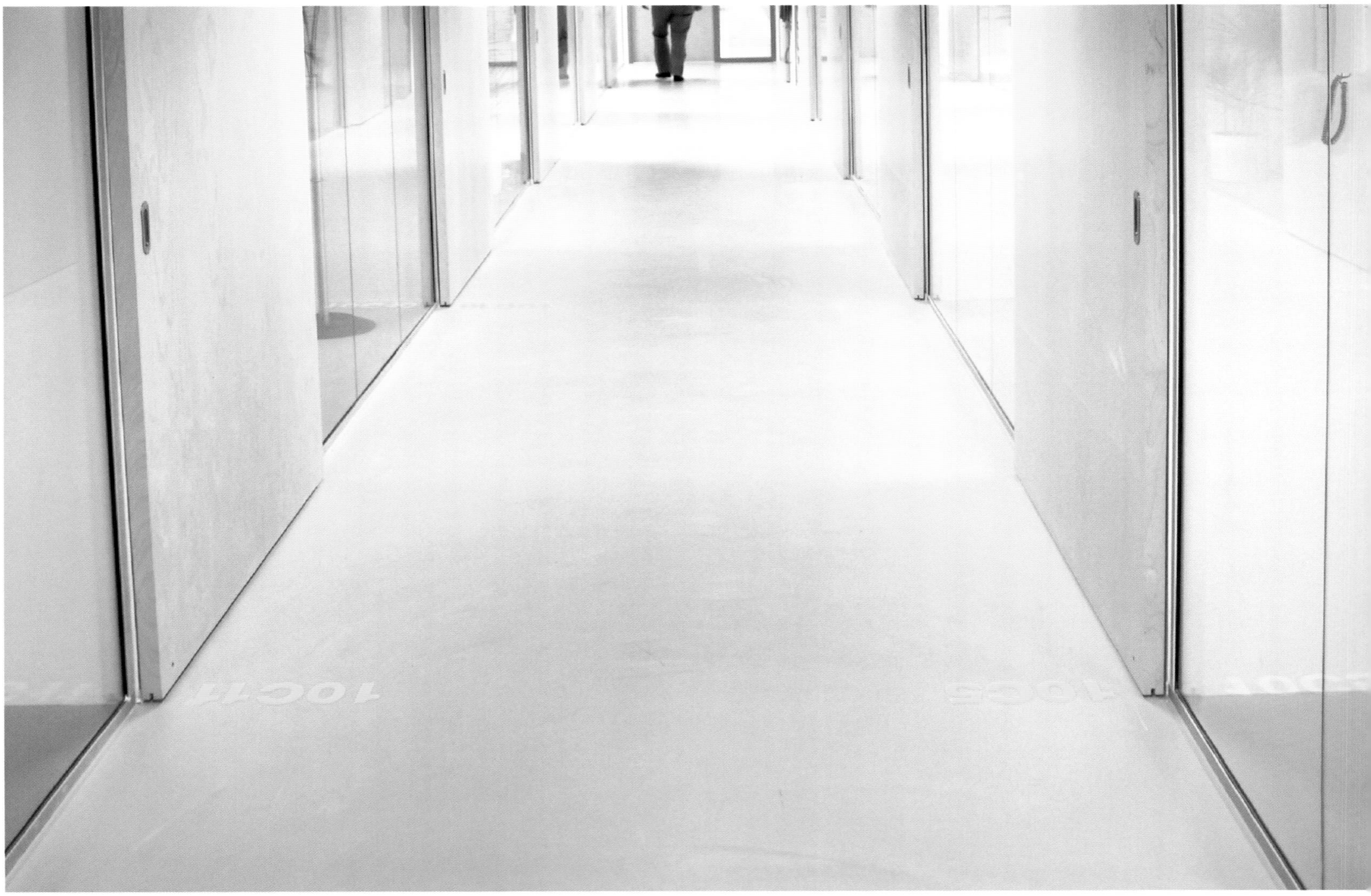

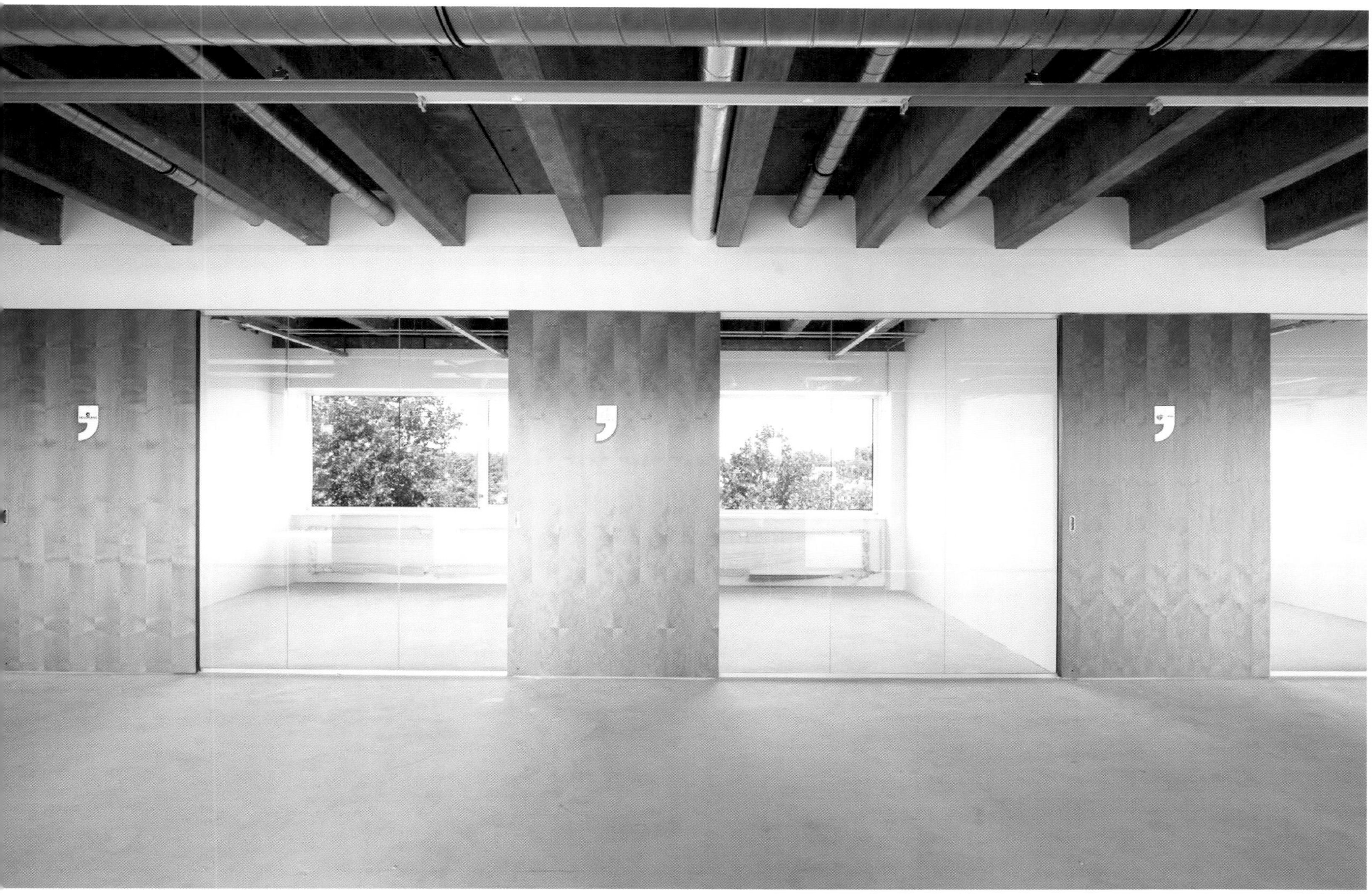

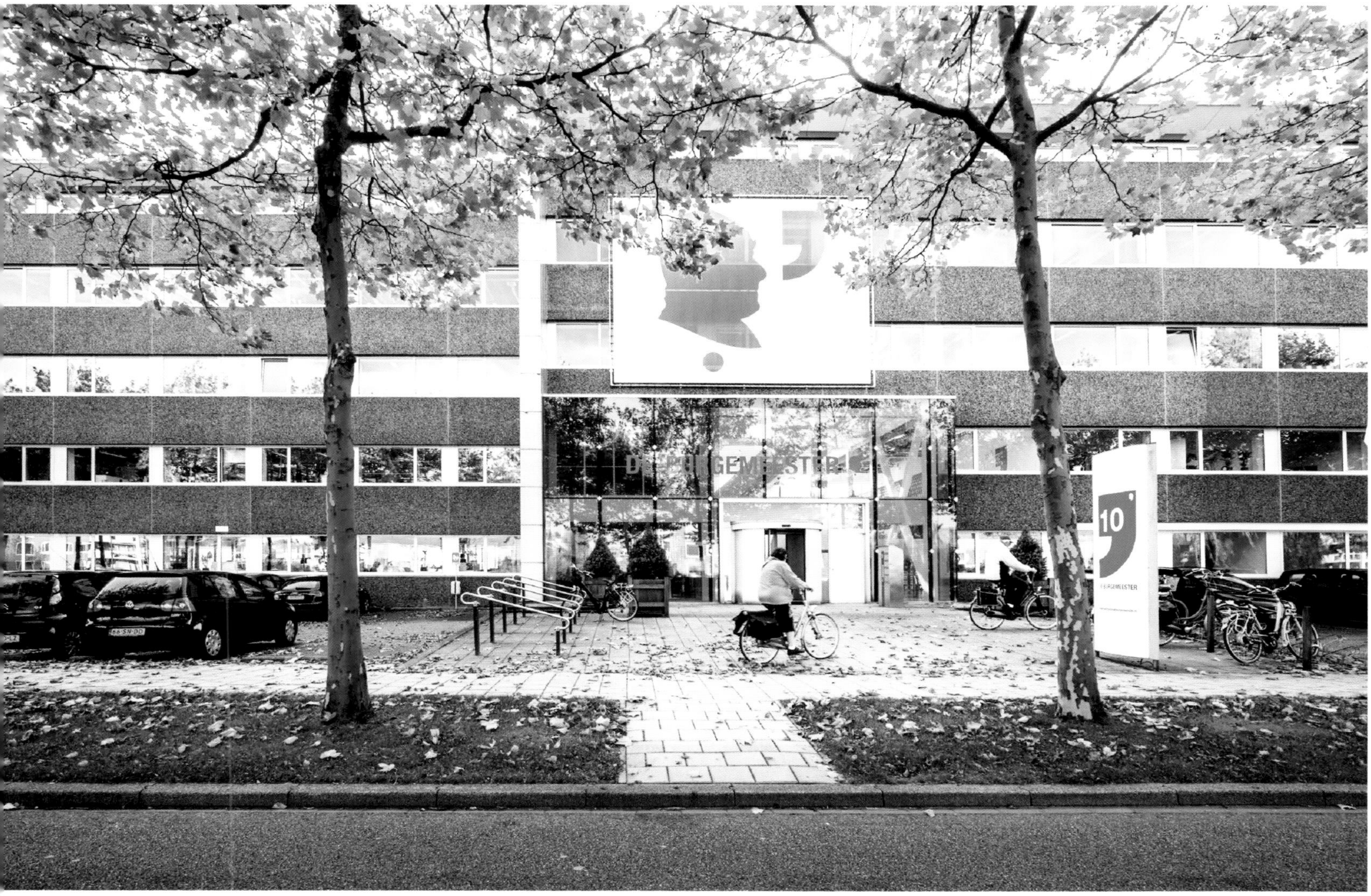
10

DRAFT Tokyo Office

东京 DRAFT 办公室

DESIGNER
Taiju Yamashita, Saeko Takahashi
DESIGN COMPANY
DRAFT Inc.
LOCATION
Tokyo, Japan
AREA
360 m²
PHOTOGRAPHER
Kaoru Fukui

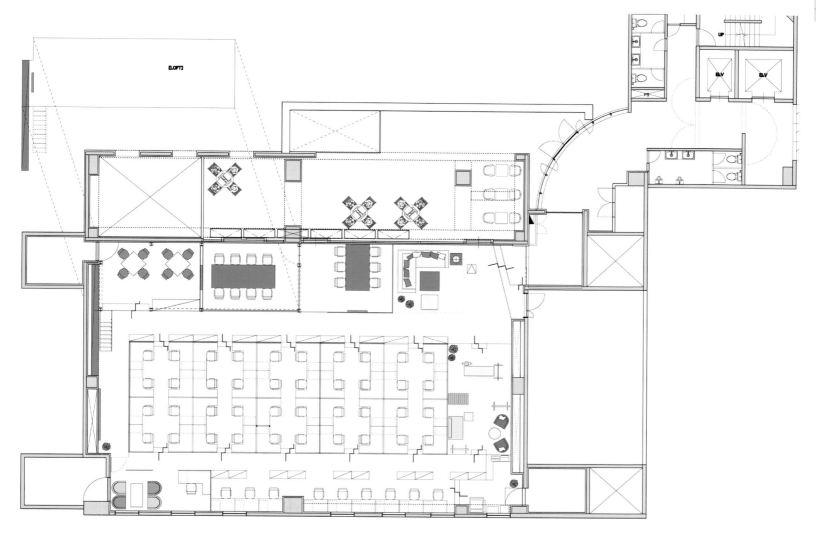

DRAFT head office is occupying the top floor of the building in Shinjuku, the center business district in Tokyo. The design team transformed previous ordinary office space into bright, pleasant, and clean large design studio. The entrance space is materialized through an industrial form with white painted plumbing. Open floor with 4.6 m high ceiling gives fresh impressions to visitors through the entrance space.

The original concept of this office was "Spontaneous Communication". Workers can hold private in cubicle office layout, but for team project such as interior design, it is necessary to build open space so that people can interact and collaborate each other everywhere in the office.

To realize both individual private space and sense of open spaciousness, the designers set square boxes into the working area that can accommodate four seats and desks which are also designed for this office by them. There are two stairs, gap (about 380 mm) between this box space and aisle in order to expand view and make communication easy from hallway. Conversely, it allows people inside the box keeping concentration to work by blocking view.

Four eye-catching big monitors on the wooden wall at the back of the room are used for conferences with other branches. Since every employee has iPhone and Apple notebook or desktop, they often use these monitors to mirror FaceTime or Skype display through Wi-Fi with Osaka, Cebu, and Shanghai staff. The meeting room is another distinctive feature. Besides providing high-level soundproof, this wall covering of the room works as a whiteboard. In addition, at the backside of the meeting room, the designers installed an unframed whiteboard on the wall. The whole office space is designed to encourage spontaneous communication and impromptu meetings.

DRAFT办公室总部位于东京中央商务区新宿区的一栋大楼的顶层。设计团队将其从原先普通的办公空间打造成一个明亮、舒适、干净的大型设计工作室。入口空间装饰有漆成白色的工业管道。穿过它，可以看见开放式的楼面布局，4.6 m高的天花板，让到访者眼前一亮。

这家办公室的最初设计概念是"自然沟通"。员工可以在隔间办公室布局中享受私人空间，但是如果碰到团队项目，例如，室内设计，就需要设计一个开阔的空间方便员工们与办公室内的任何人交流与合作。

为了实现既有私人空间又有宽敞的公共空间的目的，设计师们在工作区内安置了方形隔间，每个隔间可以容纳四个由他们为办公室设计的座椅和桌子。办公室工作区和过道之间设计有高约380 mm的两级台阶，可以让员工扩大视野，更方便地与过道另一边的员工交流。相反地，隔间内的员工则因为视线的阻隔更能集中精力工作。

房间后面的木质墙上安装了四个引人注目的大显示器，用来与公司的其他分支机构开视频会议。因为所有的员工都使用苹果手机、苹果笔记本或苹果台式电脑，他们经常通过这些监控器连接无线网使用FaceTime或者Skype与大阪、宿务岛和上海的员工们视频。会议室是另一大特色，不仅拥有高水平的隔音效果，房间内还有可以用来做白板的玻璃墙。除此之外，会议室的后墙还安装了无边框白板。整个办公室的设计理念就是方便自然的交流沟通以及召开临时会议。

Google Campus

谷歌总部园区

DESIGNER
Shaun Fernandes, Markus Nonn
DESIGN COMPANY
Jump Studios
M&E
Medland Metropolis
CONTRACTOR
Como

CLIENT
Google UK Ltd.
LOCATION
London, UK
AREA
1,858.06 m²
PHOTOGRAPHER
Gareth Gardner

Google Campus is a seven-storey co-working and event space in the center of London's Tech City, otherwise known as Silicon Roundabout.

The design challenge was to take an unprepossessing seven-storey office building and to create an interplay between dynamic, open, social spaces and more intimate working hubs, with flexibility to accommodate a shifting workforce and a diverse program of events.

Much of the architectural focus has been on opening up and connecting the ground and lower ground floors programmatically to play host to a series of socialized spaces, from reception and informal meeting areas to theater, cafe and workshop spaces.

By stripping back the building to its core, exposing all services, revealing the existing structure of ceiling slabs and columns and combining this with utilitarian and inexpensive materials such as linoleum and plywood, a raw aesthetic has been created not dissimilar to a garage or workshop.

This low-tech environment has then been furnished with several autonomous objects, which emanate a strong presence in the space.

In the reception visitors are welcomed by a reception desk partly made from multi-coloured Lego bricks – a nod to Google's founders who always had a special fondness for the Danish toy building blocks – in an otherwise unbranded environment.

A large inspiration wall made from reclaimed vegetable crates dominates the holding area. The wall can be used as shelving for books and magazine or to display objects and artefacts that help tell the story of the building and its inhabitants.

Towards the rear the holding area opens up to a large presentation room offering seats for up to 140 people. The two spaces can be subdivided by means of a bright red roller shutter which contributes to the industrial aesthetic of the environment.

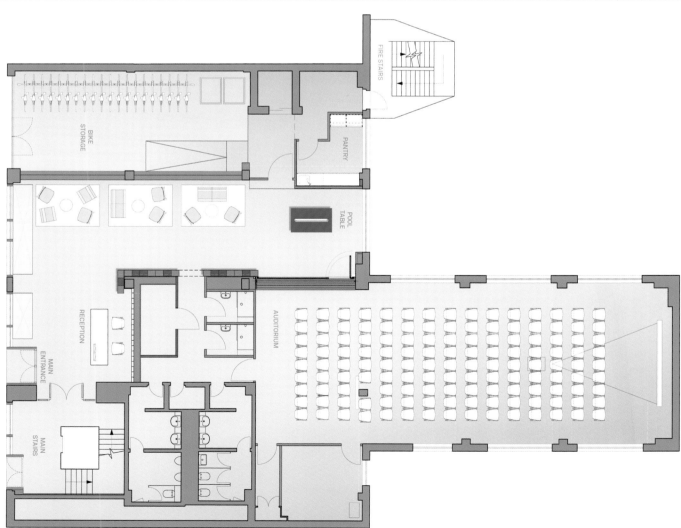

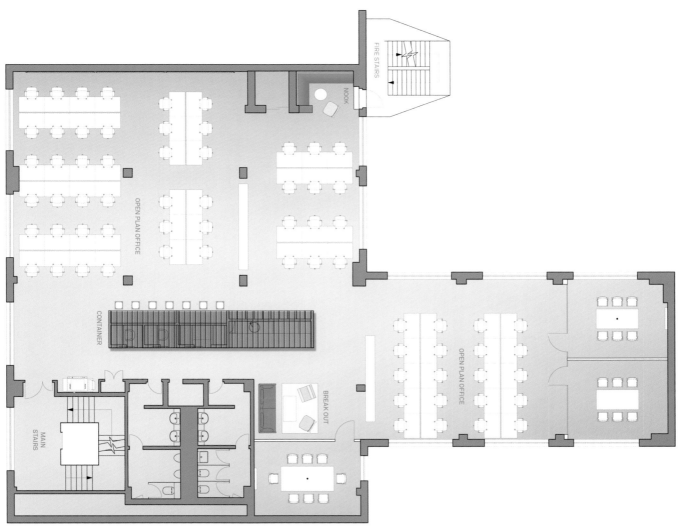

谷歌总部园区是一个七层的联合办公和承办活动的区域，位于伦敦科技城（又名小硅谷）的中心。

设计挑战在于如何在一幢毫不起眼的七层办公楼中营造出具有活力的开放氛围，使各工作部门之间的联系更加紧密，并能够与频繁的人才流动和多样的工作项目相适应。

因此，大部分的设计重点是有计划地将基层和地下层打通并合并，将其建成一系列社交活动的空间，包括接待处、非正式会议室以及礼堂、咖啡厅和研讨空间等。

设计中去掉了外层浮华，将建筑的内部结构暴露出来，因此建筑的天花和承重柱全都清晰可见。建筑材料的选用也追求实用，例如，所使用的油布、三合板等建材都十分经济实惠。这一切使得建筑具有一种类似于车库或工厂车间的粗犷美。

这个低科技含量的环境又加入了几个单独摆放的物品，在空间中散发出很强的存在感。

接待处的桌子部分由乐高积木组成——这项设计旨在致敬谷歌创始人，因为他们对这些来自丹麦的小积木块一向青睐有加。参观者们就在这样一个没有品牌标记的环境中得到接待。

储物区的主角是一面由再生蔬菜箱制成的创意墙。这面墙既可充当书架收纳书籍杂志，也可充当展览讲述园区和公司历史的艺术品。

通向后方的储物区与一个可容纳 140 人的演示厅相通。这两部分可以用明亮的红色卷帘再分成独立的空间。而作为隔断的卷帘也为设计增添了带有工业气息的美感。

Office Design of IND Architects Studio

DESIGN COMPANY
IND Architects Studio
LOCATION
Moscow, Russia
AREA
280 m²
PHOTOGRAPHER
Alexey Zarodov

IND 建筑师工作室办公室设计

When working on the project of IND Architects Studio office, the architects entrusted themselves with the task to create a contemporary creative space with a minimum of distracting elements, so that designers and architects could focus on their projects as much as possible. The office has become a business card and a reflection of the studio which has implemented a number of projects over five years of its existence.

The office has been arranged in ARTPLAY Design Center and has been decorated in the loft style. Designers have shown the benefits of a former industrial premise to their best advantage – they have kept the double-floor height area in some places, have deliberately left the concrete ceiling panels unpainted and have retained the original concrete structure – like timber shutter texture on the first floor and ceiling panels on the second floor. Exposed black-colored utilities accentuate the industrial past of the building and form a contrast to white walls of the office.

The office has been divided into two zones – one is a volumetric double-floor height area for designers and the other is a cozy space on the second floor for architects. The walls of the latter one are used for attaching the pictures and drawings of studio's current projects. In addition to two open space zones, a reception zone, a meeting room, and a coffee-point have been arranged on the first floor; an office, a leisure area, and an open meeting room for the staff may be found on the second floor. Two or three person meetings may also be held at a round table in the first-floor open space. Two separate wardrobes have been arranged for the staff and guests – one is at the entrance and the other is at the meeting room.

Grey and white are basic interior colors. Bright yellow details – such as infographics, a creatively different full-wall unicorn in the first-floor open space, and small items, like flower cache-pots, desk folders, and décor details – stand out sharply against quiet shades.

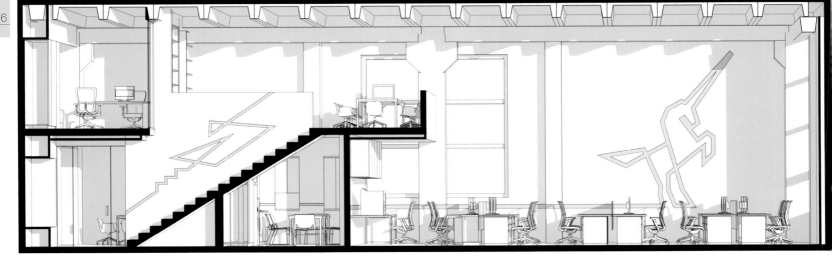

当建筑设计师们要进行自己的 IND 建筑师工作室办公室设计项目时，定位就是要创造一个具有当代特色的创意空间，最大限度地减少分散人们注意力的元素，这样设计师和建筑师们才能够最大程度地把精力投入到项目中。对于这间成立五年并已经设计过很多项目的工作室来说，办公室是整个公司的名片和形象。

这间办公室位于 ARTPLAY 设计中心，它的装潢主要以 loft 风格为主。设计师将之前的工业空间的优势进行了最好的利用，例如，在一层使用木百叶窗结构和在二层使用的天花板。这些都得益于之前的某些空间保持了双倍层高，特意留下的没有喷漆的混凝土天花板以及保持了原有空间的结构。裸露的黑色设备强调了建筑过去的工业化感觉，和办公室的白色墙面形成对比。

办公室分为两个区域：一个是为设计师建造的双倍层高的区域以及在二层为建筑师设计的舒适的空间。后者的墙面用以悬挂工作室现有项目的设计图。除了这两个开放式区域，一层还布置有接待室、会议室和咖啡室，二层有办公室、休闲区和开放式会议室。一层的开放空间还可以容纳两到三人围绕圆桌进行讨论。两个分开的储物柜是为了员工和客人准备的，一个在入口处，另外一个在会议室。

灰色和白色是室内的基础颜色。明亮的黄色细部，例如，信息图、一层开放空间中创意十足的整面墙高的独角兽图像，还有一些像花盆、文件夹和装饰的设计细节等小元素都在安静的背景下显得十分突出。

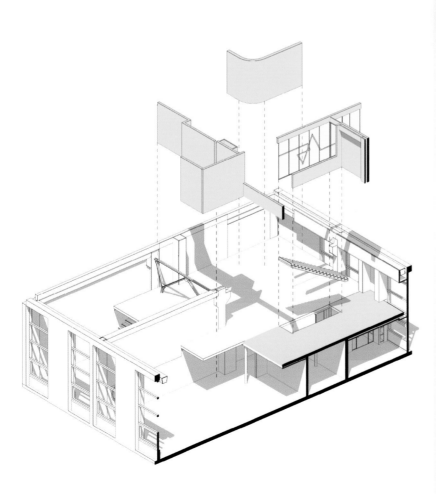

Octapharma Brewery

Octapharma 啤酒厂

DESIGN COMPANY
Joliark, White
LANDSCAPE ARCHITECT
White
CONSTRUCTION MANAGEMENT
Ebab
LOCATION
Stockholm, Sweden
AREA
7,400 m²
PHOTOGRAPHER
Torjus Dahl, Brendan Austin

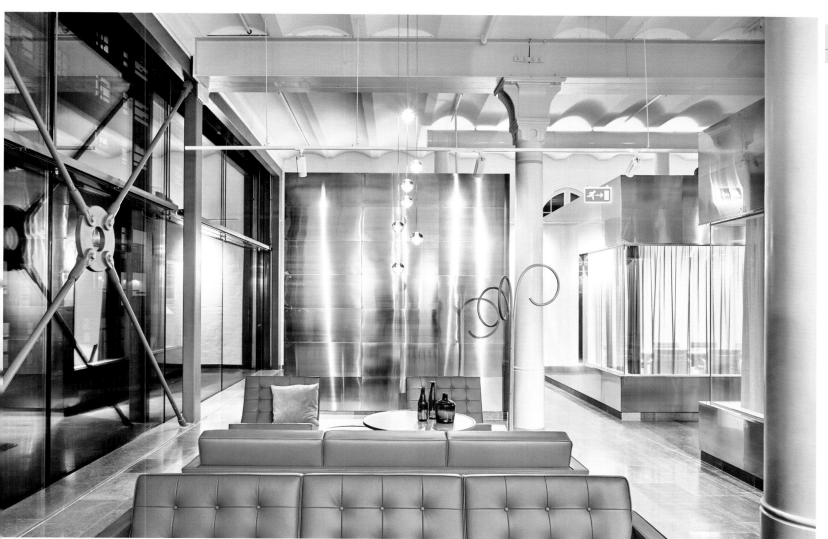

The project is a result of Octapharmas's rapid expansion and need for increased capacity, including office spaces, scientific laboratories as well as factory and production spaces within its premises in Western Kungsholmen. The former brewery, together with its surrounding facilities, was bought in 2009 from Index Estate by Octapharma and the company has since strived to improve logistics on site.

First built in the early 1890s, the brewery peaked during a 15-year period but soon faced economic difficulties and bankruptcy. The building has since been used in various ways, slowly falling into disrepair due to neglect. It has however, more recently been protected for its historic and cultural importance and is currently a listed building.

The brewery's interior design brings out the story from a time when beer was brewed in large copper vessels, and puts it in living contrast to the modern bright offices. New functions and needs are elegantly and tastefully incorporated into the old building's powerful volumes, with a system of exposed wooden beams and plastered brickwalls. Materials and furniture are carefully chosen to harmonize with the character of the building, as well as the core of Octapharma: Swedish, family-owned and global.

Copper, limestone and pine floors, combined with timeless furniture classics, create a calm and solid whole. Everything is thoroughly designed and elaborately selected, from the embossed leather desk pads to the copper-clad boxes that function as meeting rooms. In the Bistro, perhaps the brewery's most beautiful room, one sits down under the tree crowns of wooden beams to enjoy locally produced and carefully cooked meals. Octapharma's new Nordic headquarters in the brewery meets the global business, with focus on life, people and its Swedish roots.

该项目是由于 Octapharma 公司快速扩张以及满足增加产能需求而进行的，位于国王岛西部的经营场所，包括办公空间、科学实验室以及工厂和生产场地的空间。2009 年 Octapharma 公司从 Index Estate 公司手中收购了之前的啤酒厂连同其周围设施，并从此致力于改进现场物流水平。

初建于 19 世纪 90 年代早期，这个啤酒厂在 15 年内发展到顶峰，但是不久就遭遇经济困难并破产。此后，这栋楼被用于各种用途，逐渐因为疏忽而失修了。然而，最近却因为其历史与文化的重要性而受到保护，现已成为文物保护对象而登记在册。

该啤酒厂的内部设计带出那时用铜制容器发酵啤酒的故事，它与现代光鲜亮丽的办公室形成鲜明的对比。新的功能和需要优雅地被纳入旧楼的高大楼体中，与外露的木梁和灰砖墙融为一体。为了与建筑楼的风格相协调，材料和家具经过了精心挑选，也与 Octapharma 公司的核心——瑞典式、家族式和全球性——相协调。

铜、石灰石和松木地板，加上永恒的经典家具，营造出平静而牢固的一个整体。一切都经过精心设计和选择，从压花革桌垫到可充当会议室的带有铜面板的房间。小酒馆也许是啤酒厂最漂亮的房间，人们坐在冠状木梁之下享受当地特产和精心烹制的饭菜。在啤酒厂内的 Octapharma 公司的新北欧总部能满足全球的业务需求，并将焦点集中于生活、居民以及瑞典的根源。

Office IMd Rotterdam

鹿特丹 IMd 公司办公室

DESIGN COMPANY
Ector Hoogstad Architecten
PROJECT TEAM
Joost Ector, Max Pape, Chris Arts, Markus Clarijs, Hetty Mommersteeg, Arja Hoogstad, Paul Sanders, Roel Wilderyanck, Ridwan Tehupelasury
CONTRACTOR
De Combi, The Hague
LOCATION
Rotterdam, The Netherlands
AREA
2,014 m²
PHOTOGRAPHER
Petra Appelhof, Ossip van Duivenbode

Recycling is a big issue in the Netherlands today. A large proportion of the building stock is vacant, awaiting renovation or re-allocation, including premises with unsuspected qualities just waiting for people with initiative who can spot this potential. So too this steel plant on Rotterdam's Piekstraat; not an obvious location for an office, but enjoying a unique position with views over the river Maas. What made the building attractive to IMd was the vast space, dominated by an imposing steel structure.

Renovation of the existing shell of the building soon proved an unrealistic option, in both the technical and financial sense. Finally, a strategy was chosen whereby all the work areas were created on two storeys in air-conditioned zones against the closed end walls. From there, they looked back into the hall, in which pavilions with conference areas were created, interlinked by footbridges and different types of stairs.

Everything was already there, such as the steel skeleton, the concrete floors and the masonry on the facade. New additions were made using a limited number of materials which were new, but which were very much in keeping with the industrial atmosphere; rough wood for stairs, clear glass and sheeting of transparent plastic. This sheeting made the new walls nicely diffused, and even slightly "absent". The consistent use of one color – bright yellow – united the whole even more.

The clients are more than satisfied: "If recycling is really well done, the final quality will be better than that of new build. That's the motto in our work, but we are experiencing that now for ourselves, in our own office. With the contrast between the new and the existing, EHA introduced a sort of spatial 'tension', which would be impossible in the new build. It is very difficult to express what it means for the working atmosphere. But that is more different than a standard office, better and even more stimulating, and is something we experience on a daily basis."

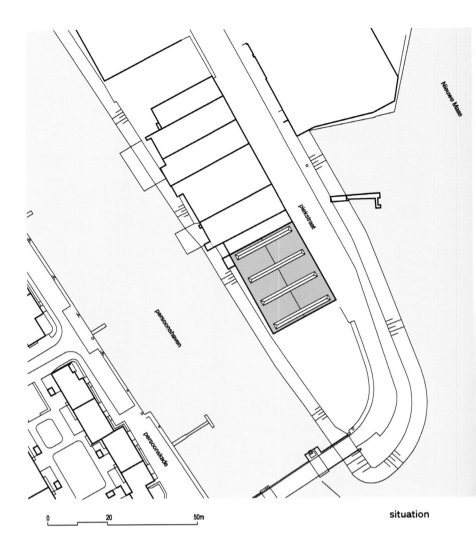

situation

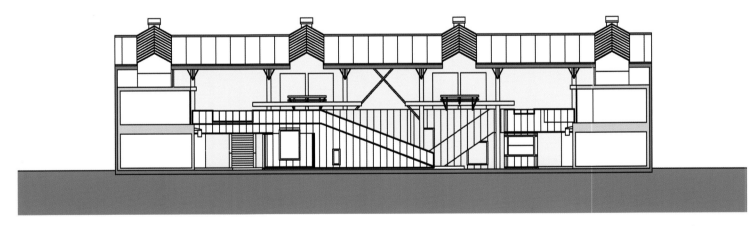

L-section

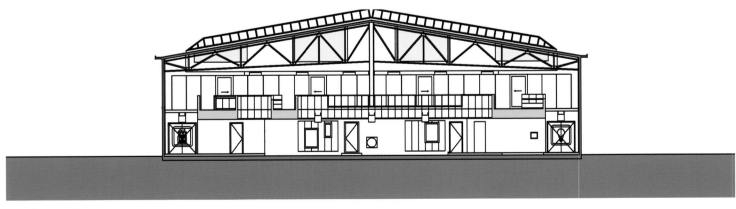

cross-section

1	entrance	
2	reception	
3	cloakroom	
4	waiting room	
5	stairs	
6	picnic tables	
7	kitchen	
8	reproduction	
9	small meeting room	
10	large meeting room	
11	administration	
12	archives	
13	copy	
14	toilets	
15	disabled toilet	
16	terrace	
17	storage	
18	incubator workspace	
19	multifunctional workspace	
20	bike park	
21	boardroom	
22	consulting room	
23	open workspace	
24	elevator	
25	server	
26	pantry	
27	meeting point	
28	individual workstation	
29	group workstation	
30	lounge	
31	library	
32	chaise longue	
33	technique	
34	vide	
35	container	
36	atrium	
37	bridge	
38	plateau	
39	trees	
40	anteroom	

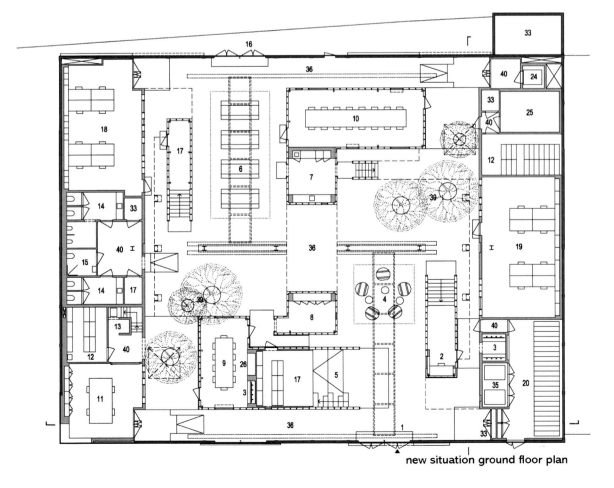

new situation ground floor plan

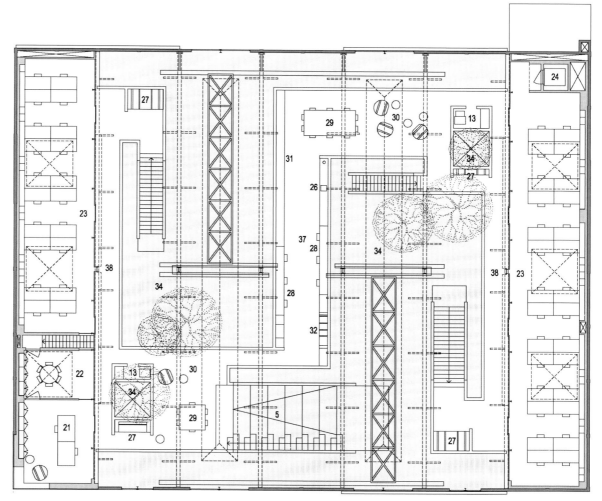

new situation second floor plan

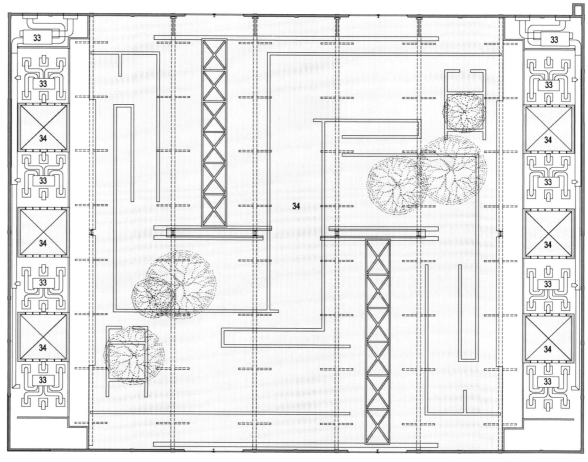

new situation third floor plan

现如今，再循环在荷兰是一件大事。大量的空闲建筑正等待着有创造力的人们来发现其未知的潜能，对其进行翻修和再分配。这家地处鹿特丹皮克斯瑞克区的钢铁厂正是如此；它的地理位置并非是做办公室的第一选择，但却拥有可饱览马斯河美景的独特位置。而对于IMd公司而言，它的引人之处在于广阔的空间和壮观的钢铁结构。

出于技术和财力方面的考虑，对建筑现有外立面进行大型改造的想法不久被证明有些不切实际。最终，确定的方案是靠端墙在空调区内修建两层办公区。从那里回头看大厅，在大厅内建造有带会议区的空间，最后用人行天桥和不同类型的楼梯使办公区和大厅相连。

设计中大量保留建筑原型，比如，钢铁骨架、混凝土地板和砖石立面。新增加的部分使用了如粗木台阶、透明玻璃和透明塑料薄膜等有限的几种新材料，但在风格上却都与原有的工业感保持一致。这种塑料薄膜对新的墙体起到了很好的隔断作用，甚至有那么一点"不存在"的感觉。同时，明黄色的重复使用又为空间增加了统一感。

客户对此十分满意，他们表示："如果再循环确实做得好，最后的质量会好过新建的。这正是我们的工作格言，但现在在自己的办公室里体验到这一点。在新与旧的强烈对比下，EHA营造出了一种'空间张力'，这在新建筑中几乎是不可能实现的。很难说这种张力对于工作氛围意味着什么，但与标准办公室相比，它更是不同，这里的氛围更好，也更加刺激，而我们每天都在体验着。

NeueHouse

新空间

DESIGN COMPANY
Rockwell Group
CLIENT
NeueHouse
LOCATION
New York, New York, U.S.A.
AREA
3,251.60 m²
PHOTOGRAPHER
Eric Laignel

NeueHouse occupies five floors of a century-old building originally designed for light manufacturing industries that later served as the location for Tepper Galleries, a well-known Manhattan auction house. The space includes a cellar level, ground floor and three upper levels of dedicated office space. Inspired by its creative members, NeueHouse hosts a rotating, curated collection of artwork by emerging international artists.

Rockwell Group conceives a work space where individual entrepreneurs and small companies can call home, and fosters the creation of real and virtual communities. Modern architectural elements have been added while retaining the original industrial feel of the building.

Members and their guests enter NeueHouse via the ground floor, a dramatic space with 18.29 m of retail frontage and 6.1 m tall ceilings. A park zone is located along the perimeter of the floor, taking advantage of the natural light from the large street-level windows. With flexible seating and furniture, including banquettes and four private phone booths, members can relax and socialize, while passersby can look in to witness the vibrant, inspirational activities taking place inside.

Floors 2 to 4 are dedicated to work space which serves as an "incubator" that can grow and expand to accommodate the members' enterprises. Members can pay an additional fee to access these floors and lease their own dedicated office. Custom, transformable "partner desks" made from plywood and steel provide permanent seating for two to four collaborators. These desks can be configured in multiple ways.

The basement level offers larger shared environments. It features an auditorium / theater with 47 seats, a radio broadcast booth that can be used for live recordings or an internet radio station, a "quiet" library, lounge, and a package room that can be used as a bar or food service. The main conference room is located on this floor and is illuminated by natural light streaming in from a glass ceiling that evokes old sidewalk vault lights made from cast iron panels fitted with glass tile. The conference room can also be transformed into a private dining room for events.

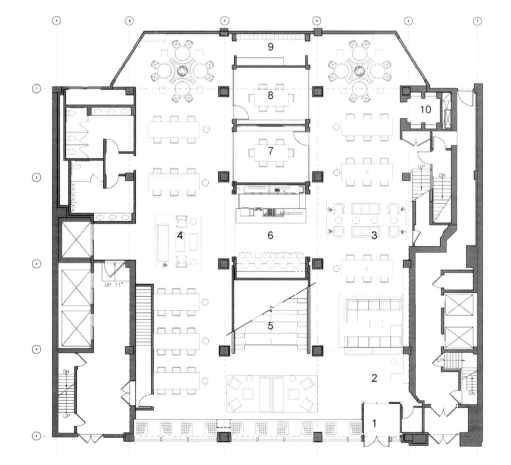

1 Entry Vestibule	6 Cafe
2 Check-In	7 Conference Room A
3 Shared Work Area	8 Conference Room B
4 Shared Work Area	9 Print Pantry
5 Seating	10 Phone Booths

NeueHouse Ground Floor

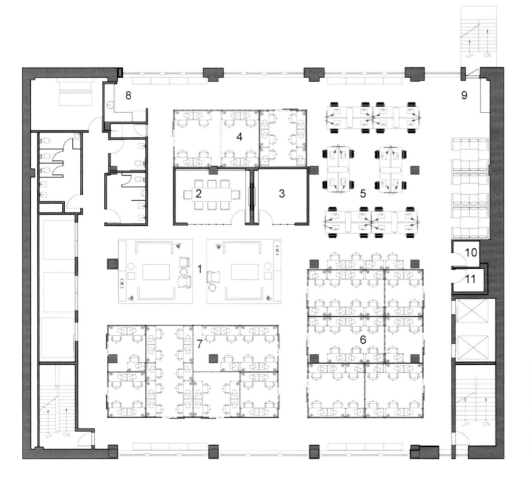

1	Open Meeting	6	Dedicated C
2	Conference Room A	7	Dedicated D
3	Conference Room B	8	Pantry
4	Dedicated A	9	Print Pantry
5	Dedicated B	10	Phone Booth 1
		11	Phone Booth 2

Neuehouse Typical Floors 2 - 4

新空间占据了一座拥有百年历史的建筑的五层空间，这里起初是为灯具制造行业设计的，此后用于曼哈顿著名拍卖行泰珀艺廊。其内部空间包括地下一层、一层及其他三层办公专区。深受其极富创造力的会员的影响，新空间收藏了大量新兴国际艺术家的艺术作品。

洛克威尔集团的设计初衷是为企业家和小型公司打造具有"家"般氛围的工作环境，并以此来推动现实和虚拟社区的建设。在保持建筑原有工业感觉的同时，现代的建筑元素也被添加到了设计中。

会员及其客人们通过一层进入新空间。一层是一个引人注目的空间，高 6.1 m，有着 18.29 m 长的临街商铺。停车场位于这一层的外沿，自然光线通过巨大的临街窗口为其照明。空间内灵活摆放着座椅和家具，还有长沙发和四个电话亭，会员可以在此放松和沟通。而与此同时，路过的行人也可以看到这里生机勃勃的活动景象。

二到四层的办公专区是会员企业的"孵化器"，其空间可以进行扩展来容纳会员企业。会员可以支付额外费用，来获得办公楼层的访问权限并租赁自己的专属办公区域。特殊定制的、可变化的"合伙人桌"由三合板和钢制成，可供两到四名经营者落座。这些桌子能够以多种形式组合。

地下室层提供了更大的共享环境。其内设一个 47 座的礼堂、一个可以进行实况录音或互联网广播的电台直播间、一个"幽静的"图书馆、一个休息室以及一个可以用作酒吧或提供食物的包装间。主会议室也位于这一层，自然光线透过玻璃顶棚射入，为房间提供照明。这种顶棚会使人想起旧时人行道上带有玻璃罩的铸铁路面采光窗。有特殊活动时，会议室也可作私人餐厅。

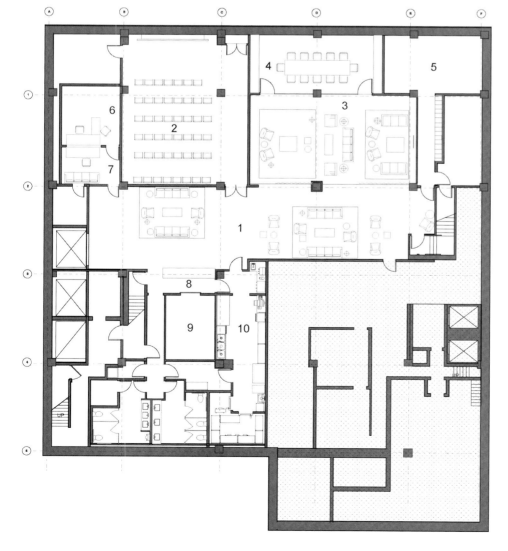

1 Prefunction Room	6 Radio Station
2 Training Room	7 Control Room
3 Lounge	8 Package IT Counter
4 Conference	9 Package Room
5 Office	10 Pantry & Storage

NeueHouse Cellar Floor

Yandex Stroganov Office

斯特罗加诺夫 Yandex 办公区

DESIGNER
Arseniy Borisenko, Peter Zaytsev
DESIGN COMPANY
za bor architects
PROJECT COORDINATOR
Nadezhda Rozhanskaya
CLIENT
Yandex
LOCATION
Moscow, Russia
AREA
5,800 m²
PHOTOGRAPHER
Maria Turynkina, Dmitry Kulinevich

The main projects in za bor architects' portfolio are held by offices of IT-companies. It has a lot to do with a pretty informal and creative atmosphere that these firms are willing to build up for their employees, because working environment is one of the key factors that affect the company's attraction. It is worth to note that Yandex – the largest IT-company in Russia, and one of the world's leaders in this field, has been intrusting their offices to za bor architects for six years already. Today there are 21 Yandex offices in 12 cities of four countries around the world, which za bor architects have developed.

Recently one more Moscow office of Yandex is opened in Stroganov building in Krasnaya Roza 1875 business quarter. This reconstructed building is full of columns and interstorey premises, which influenced the interiors a lot. The client, as usually, wants to see a happy and comfortable interior that will hold a large number of specialists.

The first three floors have the following common elements of all Yandex offices, as open communication lines on the ceiling, unique ceiling lights in complex geometrical boxes, and compound flowerpots with flowers dragging on to the ceiling. Alcove sofas by Vitra are used as bright color spots, and places for informal communication. Wall finishing includes traditionally industrial carpet, marker covering and cork.

The fourth and fifth floors are constructed in a totally different style. You may only notice two signature elements of za bor architects here – large meeting rooms and the bold colors – architects call them "bathyscaphes", and employees named them "orange" and "tomato" due to their colors.

Such difference in decoration is determined with very complex construction elements, level differences in the building (the ceiling height varies from 2 m to 6 m), balconies, and beams that are left from the previous tenants. Nevertheless, here people can see new colors, partition walls and flooring. Here, in these neutral grey-white interiors, rather than elsewhere, there are many workplaces completed with Herman Miller systems, and the largest open-spaces.

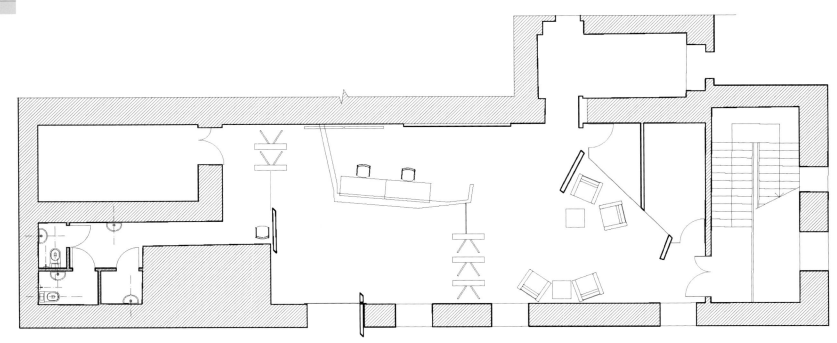

1st Floor Plan

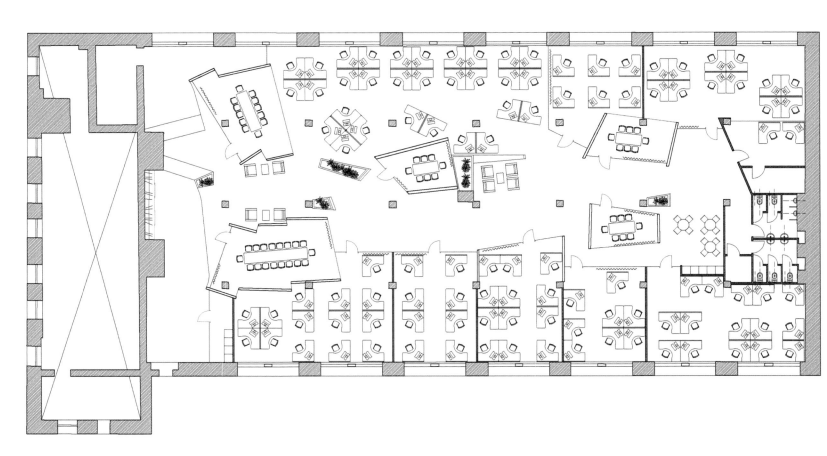

2nd Floor Plan

za bor 建筑事务所主要的设计项目都是信息技术产业公司的办公室。由于对于这些公司而言，工作环境是影响其吸引力的关键因素之一，所以他们想为员工提供随意且富有创造性的氛围，这从设计上来说有很多工作可做。值得一提的是，Yandex 公司是俄罗斯最大的信息技术产业公司，也是全球信息技术领域的领导者之一，已经连续六年将其办公室设计委托给 za bor 建筑事务所进行设计。现今，全球 4 个国家 12 座城市中共有 21 个 Yandex 办公室，它们都由 za bor 建筑事务所设计完成。

最近，又一个 Yandex 办公室落户于莫斯科卡拉斯拉雅罗扎 1875 商务区的斯特罗加诺夫大楼内。这座重新改建的建筑内有很多圆柱以及跨楼层的办公室，对内部影响很大。同以往一样，客户想要看到一个令人愉悦、舒适，并可以容纳许多专业人士的室内空间。

前三层拥有以下所有 Yandex 办公区共有的元素：天花板上开放的通信线路、独特的几何形状的盒子灯以及"栽"在天花板上的盆中花。设计师们选用威达牌凹状沙发为办公区增加亮色调，并为非正式交流提供场所。墙体装饰使用传统工艺编织毯、指示牌遮盖物和绳索。

四层与五层采用了与前三层完全不同的建筑风格。在这里，你也许只能发现 za bor 建筑事务所的两个识别特征——大型会议室和醒目的色彩——建筑师们称会议室为"深海潜艇"，而由于它们的颜色，员工们把它们叫作"橘子"和"西红柿"。

装饰上的这些差别是由复杂的建筑元素、建筑的楼层差异（层高由 2 m 至 6 m 不等）、阳台以及由前租客留下的横梁所决定的。尽管如此，在这里，人们可以看到新色彩、隔墙以及地面。在这里，在中性色彩的灰白室内空间中，与其他地方相比，人们可以看到许多工作区采用了赫曼·米勒体系，并有着最大化的开放空间。

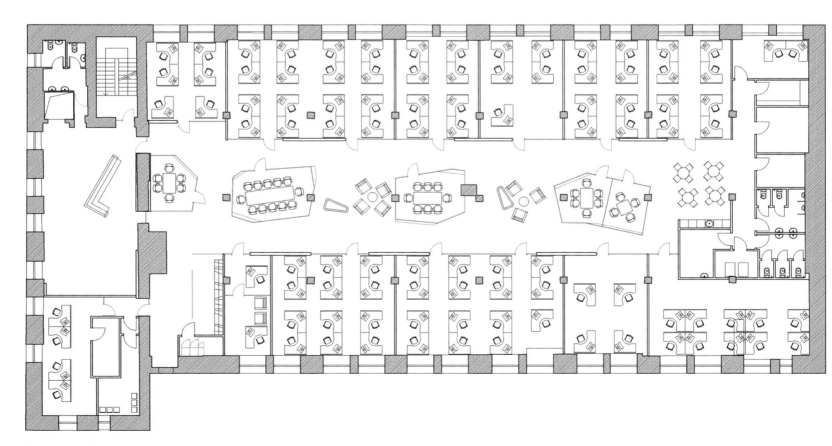

3rd Floor Plan

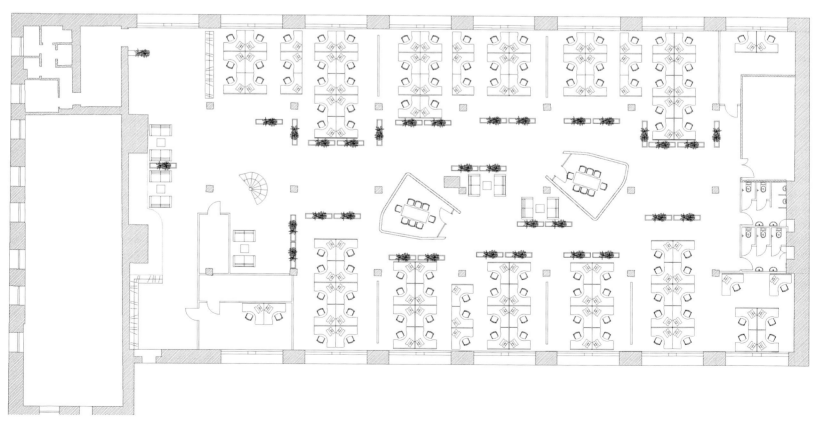

4th Floor Plan

Coca-Cola

可口可乐办公室

DESIGN COMPANY
Arquitectura e Interiores SAS
LOCATION
Bogotá, Colombia
AREA
2,400 m²
PHOTOGRAPHER
Juan Fernando Castro

This project aims to give a better quality of life for their employees as well as reflect the essence of the brand. It is designed to be a fresh and pleasant, optimistic and cheerful space. The project which developed in two plants begins with an entrance hall with handmade wood ceiling representing Coca-Cola's logo waves. The palette is neutral. It brings out the white tones. Black, beige and wood allow a perfect balance with red distributed in other design elements.

As the idea behind the design was to inspire employees, Coca-Cola was looking for a project that through design elements staff could find a creative and enjoyable space. A red container, a rustic wooden box, a red bottle, old posters and other items that evoke the beginnings of the brand served as inspiration and resulted in materials or architectural elements.

The office represented general trends in new ways of working: open offices and communication spaces. The project received the LEED (Leadership in Energy & Environmental Design) Certification in 2015.

这个项目旨在提高员工生活质量以及体现其品牌的精髓。其设计追求清新怡人、积极活泼。该项目在两个工厂进行，首先是门厅手工制作木质的天花板展示出可口可乐的标志波纹。空间的色彩呈中性，并突出白色色调。黑色、米色和木色与其他设计元素中的红色达到一种完美的平衡。

由于设计背后的意图是激励员工，所以可口可乐公司一直在寻找一个项目，使员工通过设计元素发现一个既有创意又有趣的空间。一个红色的集装箱、一个质朴的木箱、一个红色瓶子和旧海报等物品，这些最初成为品牌灵感来源的物品，最后也成为材料或建筑元素。

办公室体现了新型工作方式的总体趋势——开放式办公室和交流空间。在2015年，该项目获美国绿色建筑评估体系LEED认证。

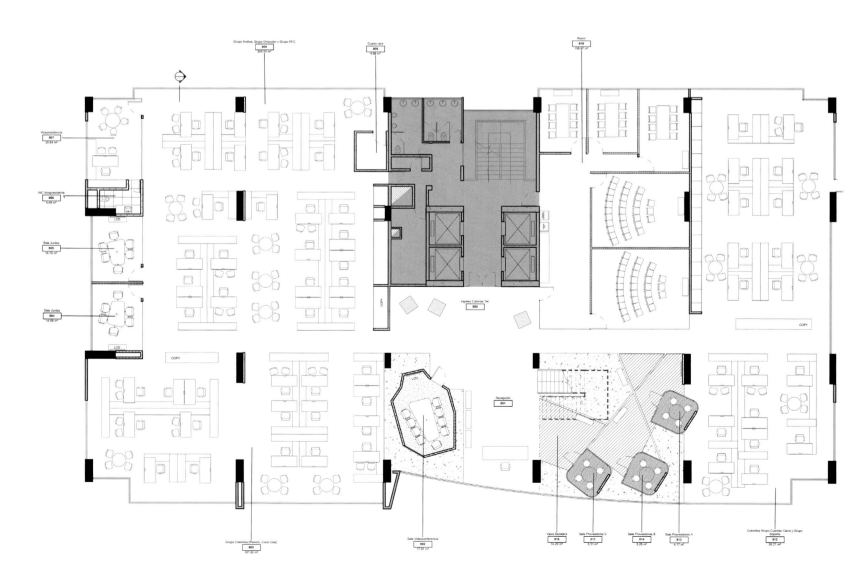

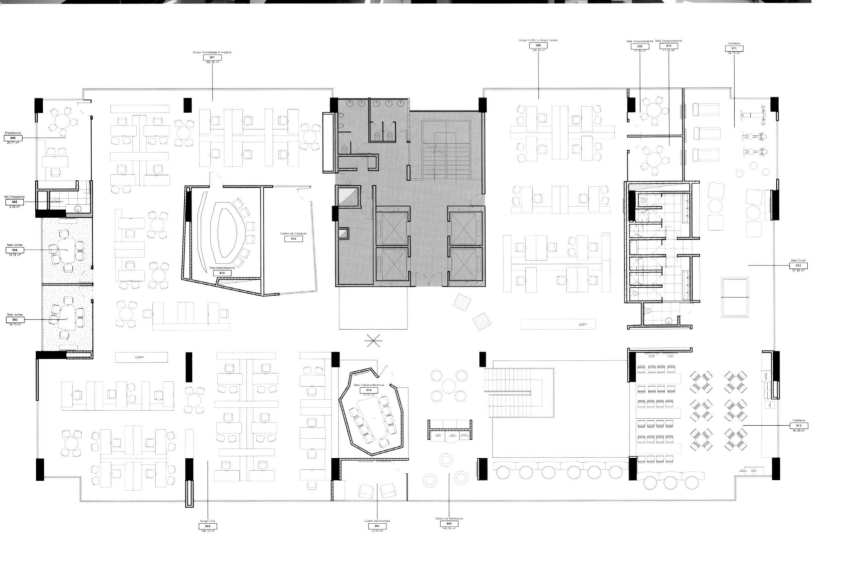

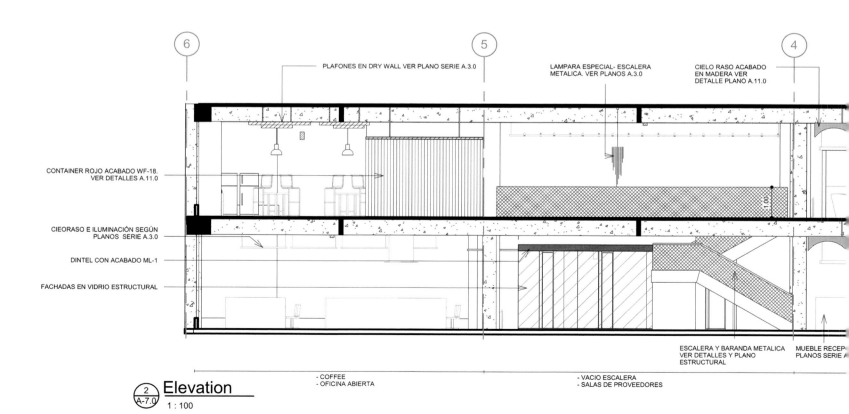

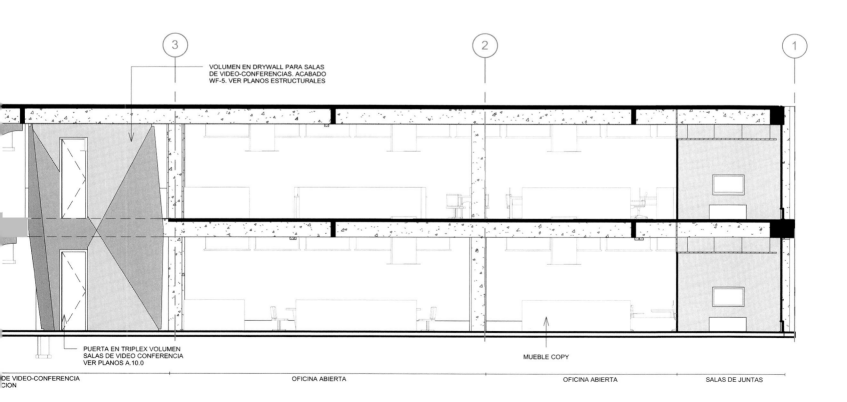

VOLUMEN EN DRYWALL PARA SALAS DE VIDEO-CONFERENCIAS. ACABADO WF-5. VER PLANOS ESTRUCTURALES

PUERTA EN TRIPLEX VOLUMEN SALAS DE VIDEO CONFERENCIA VER PLANOS A.10.0

MUEBLE COPY

DE VIDEO-CONFERENCIA | OFICINA ABIERTA | OFICINA ABIERTA | SALAS DE JUNTAS

Zendesk San Francisco Headquarters

Zendesk 旧金山总部

DESIGN COMPANY
Design Blitz
LOCATION
San Francisco, California, U.S.A.
AREA
6,968 m²
PHOTOGRAPHER
Bruce Damonte

Zendesk, a customer service software provider, was one of the first tech companies to move to San Francisco's Mid-Market neighborhood to take advantage of the city's tax incentive program, and the first to sign a community benefits agreement. After experiencing large-scale growth, it was only natural for the company to seek an additional space in the neighborhood they called home. Zendesk's new office marked its second building on Market Street, creating an urban campus that assisted in the revival of the long-neglected Mid-Market neighborhood. The new building not only served as a headquarters for the growing Zendesk team, but also opened its doors as a communal hub for local businesses and residents.

The qualities of being airy, humble, charming, and uncomplicated were what made up the core of Zendesk's brand attributes, and were used as guiding principles for design. Zendesk wanted the space to feel light and open, but also varied and textured. By including secluded and darker spaces, Blitz created a contrasting atmosphere that emphasized the airiness of the open office.

Organic materials were selected throughout to harmonize with the building's existing finishes and to emphasize Zendesk's humble qualities. The company's charming attributes were also evident throughout, from the ground floor devoted entirely to reception, to pops of green cheekily inserted into the calm and neutral palette. Finally, Blitz took a reductionist approach, looking for opportunities to reduce rather than add. The result was an uncomplicated yet thoughtful space that combined minimalism with warmth.

Another key design influence was the Danish concept of "hygge," which was roughly translated into "coziness". Blitz created spaces with lowered ceilings and muted lighting, finished with soft, acoustic wall coverings and natural materials to create an inviting contrast to the bright and large open office area. Custom-designed booths created cozy refuges for meetings and focused work. Wood canopies extended from the kitchens, providing additional nooks and shelters.

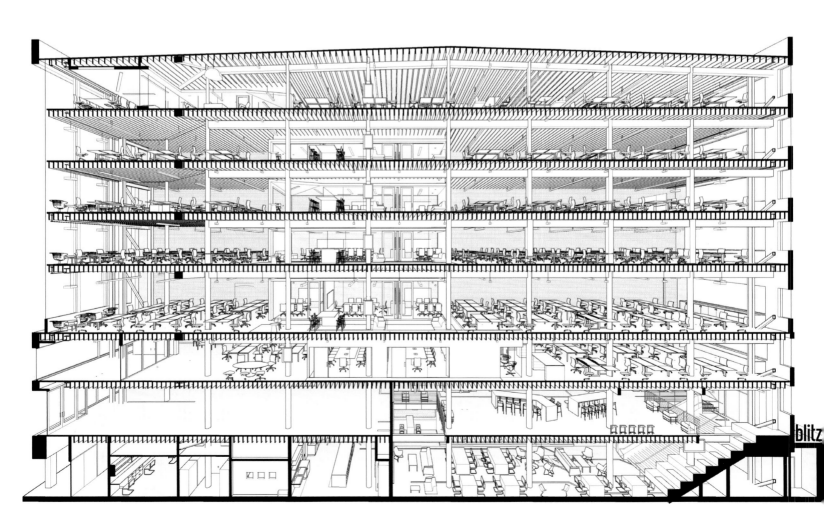

　　Zendesk是一家客户服务软件供应商，该公司是最早一批搬到旧金山中间市场区，享受税收鼓励政策的科技公司，也是第一批签订社区福利协议的科技公司。经过一段时间的快速发展之后，公司会自然而然在附近寻找一个他们称之为家的额外空间。Zendesk的新办事处位于市场街，标志着公司第二个办公场所的诞生，这座都市园区有助于重振长期被忽视的中间市场区。新办事处不仅作为日益壮大的Zendesk团队的总部，而且还用作公共中心，为当地企业和居民敞开大门。

　　通风好、朴素、迷人和简单的特点成为Zendesk品牌特质的核心，并作为其设计的指导原则。Zendesk希望使空间既敞亮，又要不尽相同且富有质感。那些黑暗隐蔽的空间也都包括在内，布利茨设计事务所打造出一种截然不同的环境，即注重开放式办公室的通风。

　　选用的有机材料与现有建筑物表面相得益彰，还凸显了Zendesk的朴素风格。公司亮点也是最显而易见的特质，从整个一楼大厅用作接人待客，到在这种平静和自然的色调中肆意使用绿色。布利茨设计事务所终究还是归于简约，设计只减不增。结果是打造出一个简约而不失人性化的空间，把极简风格和温暖感觉融为一体。

　　另外一个关键的设计导向来源于丹麦理念"hygge"，姑且将其理解为"惬意"。布利茨设计事务所采用低矮天花板配着柔和的照明，用软性隔声墙纸和天然材料收尾来打造出明亮的大型开放式办公区。另外，设计师们还专门设计了用于开会和进行重要工作的舒适小房间，木檐板从厨房延伸出来，提供了额外的角落和场所。

Valencia

瓦伦西亚舞蹈厅

DESIGN COMPANY
Dorte Mandrup Arkitekter A/S
ENGINEER
Jorgen Nielsen A/S, JJ Byg A/S
CLIENT
The Dreyer Foundation
LOCATION
Copenhagen, Denmark
AREA
2,420 m²
PHOTOGRAPHER
Torben Eskerod

The original dance pavilion Valencia, designed by H.S. Stilling was built in 1861 and was a building with a long history in the Copenhagen entertainments. It has witnessed the rapid growth of the neighborhood as the Copenhagen Ramparts were released in late 1800s, and has at first-hand felt the city's explosive growth as the detached dance pavilion in 1889. Before the thorough renovation the hall stood unnoticed, and the aim was to bring Valencia back in to the consciousness of the Copenhagen residents.

The intention of the project was to expose the building so that the unique spatial qualities were clarified, the visual connections and the spatial flow enhanced, and the individual rooms cleaned for unsettled architectural additions. The existing buildings consisted of three different building typologies; the block-housing (front building), the assembly-hall, and the rear building, which synergistically offered a large range of applications.

The front building, with its facade to the busy Vesterbrogade, is a classical Copenhagen property in 5 floors. The high-ceilinged entrance is radically changed from its current closed form to a modern minimalist, and furnished the transparent ground floor with continuous glass facades on both sides. Therefore, that the characteristic old hall placed in a new fairytale garden, is exposed to Vesterbrogade. The ground floor forms naturally as a lobby for the Association of Danish Law Firms. The first three floors of the building house the administration for the Association of Danish Law Firms, and the top 3 floors for residential apartments.

The hall is the project's central space, providing a framework for Danish Lawyers' course and conference activities. A large-scale furniture is located in the hall, which contains both the vertical relationship between basement, ground floor and first floor and a floating meeting box located in a visual open connection with the chamber. The existing walls of the hall have been maintained and renovated as gently as possible.

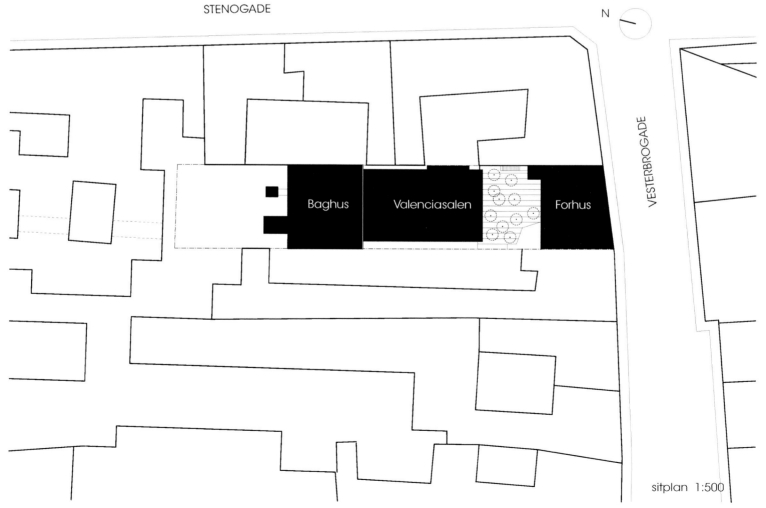

STENOGADE

Baghus | Valenciasalen | Forhus

VESTERBROGADE

sitplan 1:500

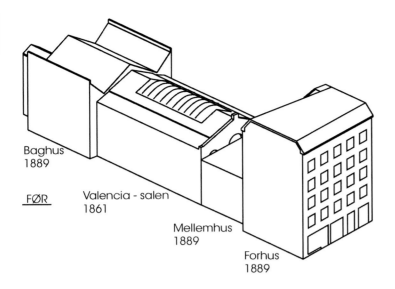

Baghus 1889
FØR
Valencia - salen 1861
Mellemhus 1889
Forhus 1889

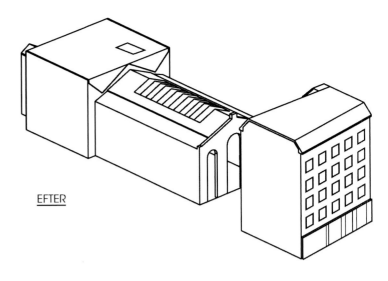

EFTER

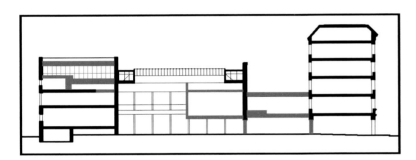

Eksisterende bygninger fritlægges.

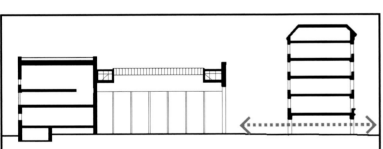

Nye elementer tilføjes.

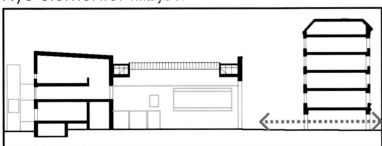

Visuelle forbindelser på tværs af bygningerne.

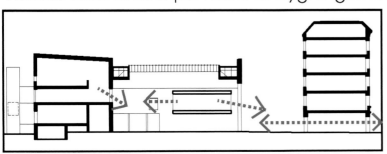

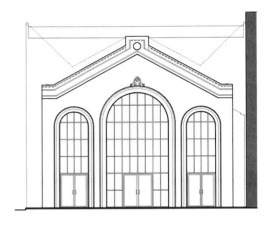

Facade 1:200

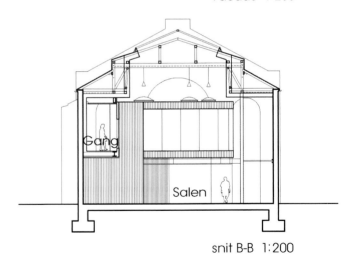

snit B-B 1:200

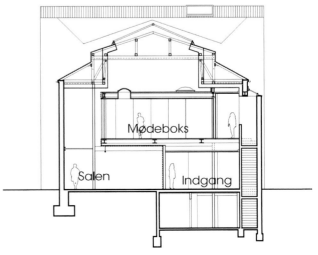

snit C-C 1:200

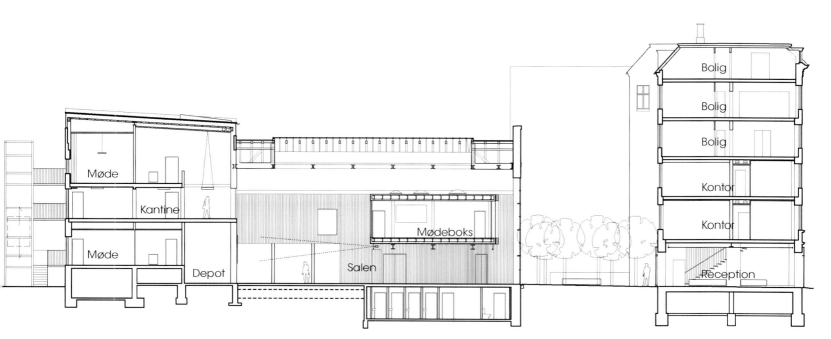

snit A-A 1:200

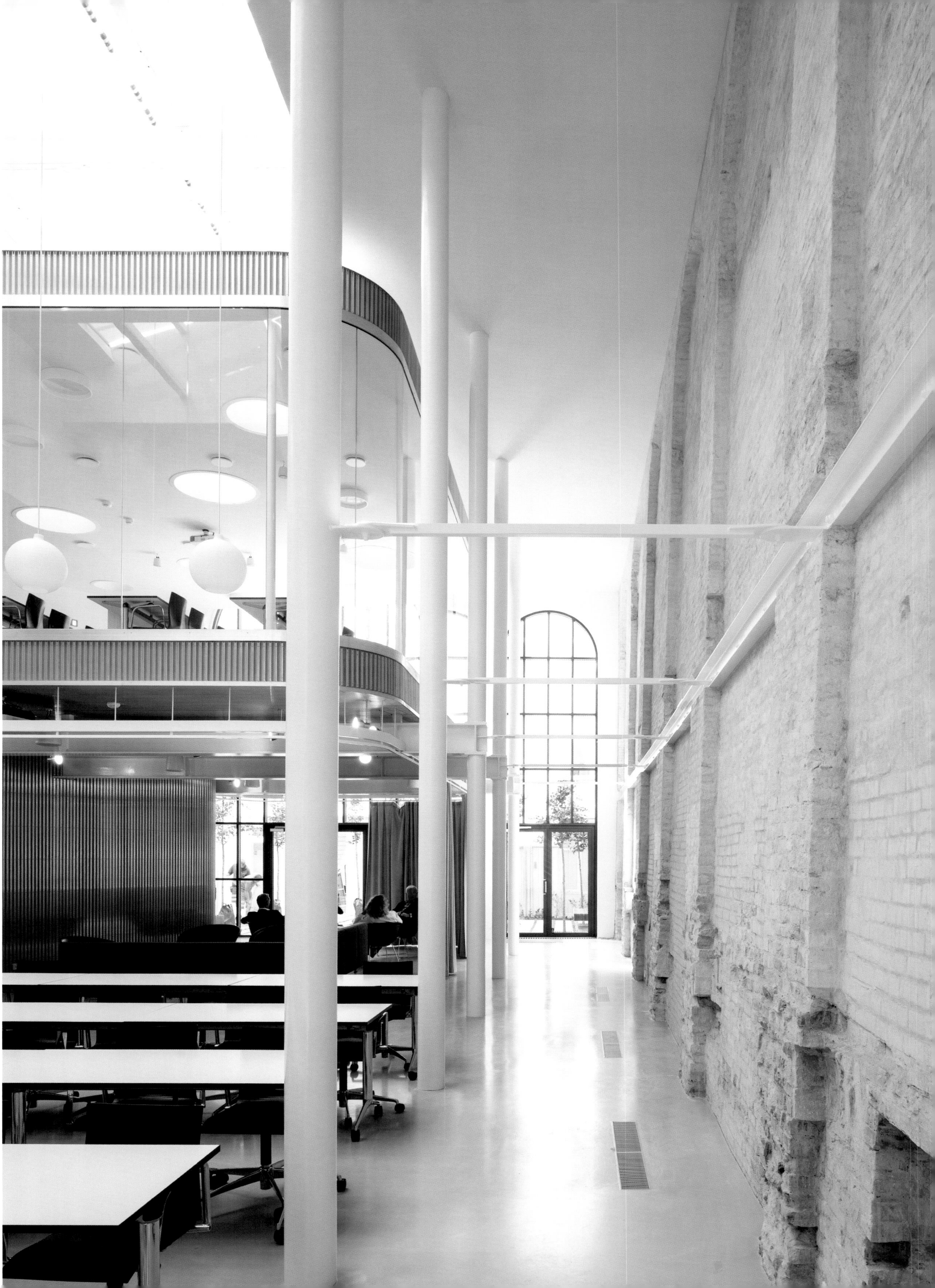

最初的瓦伦西亚舞蹈厅建于 1861 年，由 H.S. 斯蒂林设计。其历史悠久，在哥本哈根娱乐史上占有一席之地。19 世纪后期，哥本哈根城墙被拆除，自此，瓦伦西亚见证了周边的快速发展并对这种爆炸性的发展有了切身感受——1889 年，这的舞蹈厅被孤立起来。在彻底翻修之前，整座建筑无人关注，翻修的目标就是使瓦伦西亚重回哥本哈根人的视线之中。

项目的目标就是释放该建筑，以使其独特的空间品质变得清晰，视觉连接及空间流动性得到加强且独立的房间被清理干净做扩建所用，具体扩建形式还未决定。现有建筑由三部分不同的建筑类型组成：大厅区（前部建筑）、礼堂以及后部建筑，三者协同合作，功能多样。

前部建筑为古典式哥本哈根建筑，有五层高，正面临繁华的凡斯特布罗戈德大街。高顶入口完全由现有封闭式形式改造成一种具有现代极简主义风格的形式，并将通透的一层空间两侧都设置为玻璃立面。这样，这个位于一座新童话花园中的独具特色的古老大厅便完全呈现在凡斯特布罗戈德大街上。一层自然成为丹麦律师事务所协会大厅。建筑的下面三层为丹麦律师事务所协会管理处，上面三层为住宅公寓。

礼堂为项目的主体部分，为丹麦律师以及会议活动提供了活动场所。礼堂中摆放了大型家具。礼堂中还包括有连接地下室、一层和二层的纵向楼梯，以及可以俯瞰礼堂的悬空会议室。设计师们将礼堂的现存墙面保留了下来，并对其进行了小心的修复。

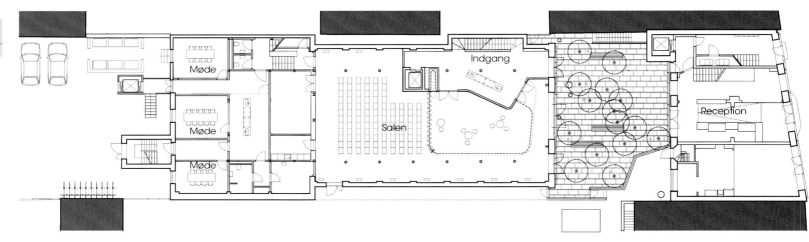

Stueplan 1:200

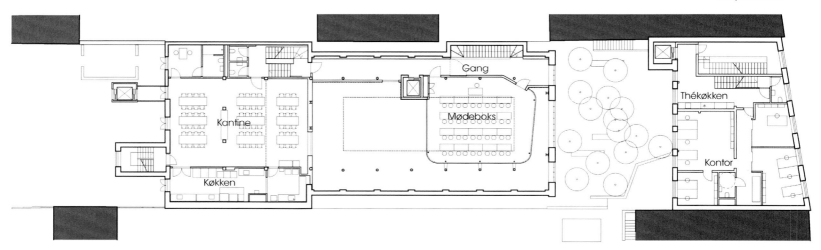

1.sal 1:200

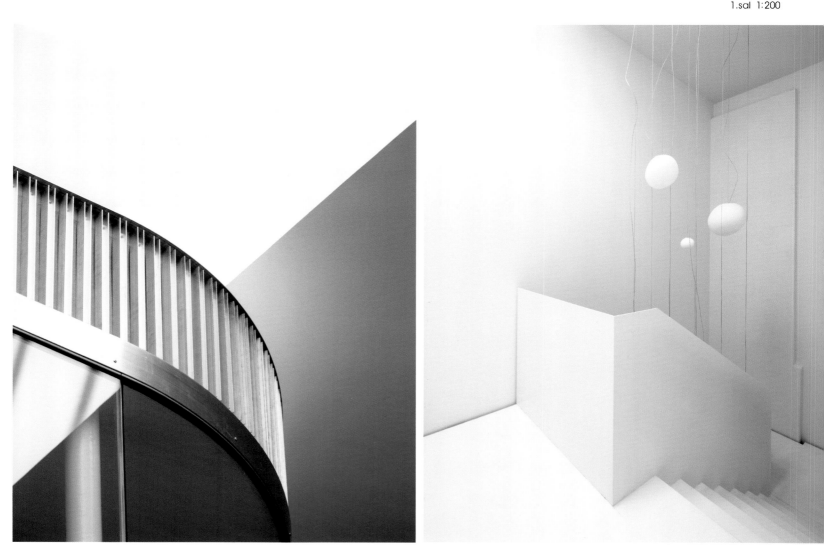

EHE Design Xiaohe Road Office

易和设计小河路办公室

DESIGNER
Ma Hui
DESIGN COMPANY
EHE Design
LOCATION
Hangzhou, Zhejiang Province, China
AREA
1,565 m²
PHOTOGRAPHER
Geng Tao

This space is use for EHE Design headquarters office. It is transformed from one of the silk reeling workshops of Grand Canal of China Tongyigong Cotton Mill. The designer keeps the original structure of cultural relics, renovates the internal pattern, not only retains the natural historical sense, but also integrates modern fashion elements. The design not only inherits the old factory building of industrial temperament, but also turns it into a fashionable and modern office space through the creative layout and design.

For designers, the space interpretation and wood structure which experienced hundreds of years are wonderful. The design respects the original space, material, historical style and features, controls the use of design language, protects cultural relics but to reach an accurate design realm. The full space character shows white, green and gray colored BOLON vinyl tile combined with hundreds of years old wood structure. Necessary illumination and air conditioning are hidden behind the white cabinet and the wood structure. This strong color impact and contrast between the old and the new create a traditional ultra-modern space.

Traditional textile workshops have north sided high windows in order to prevent silks from intensive light. For an office space, that kind of lighting is not ideal, which making light as the most difficult part of design. After careful scrutiny, the designer keeps the north sided windows, adds a bunch of fresh air system, and paints the original wood structure in white, making the space much brighter. Secondary lighting is used, which is softer and closer to natural light, making the office space more comfortable.

Grand Canal of China succeeded in application for World Heritage in June 2014. This case, as typical projects of the old industrial factory building renovation in Hangzhou's application, wins the affirmations from all walks of life with its distinctive characteristics and unique charm. It is like a pearl, which constitutes a rich human landscape of the canal together with the material culture heritages related to the canal.

本案是易和设计公司总部自用的办公空间，是由中国京杭大运河通益公纱厂的缫丝车间改造而成。设计师在保持原有文物结构不变的前提下对内部格局进行了重新的规划改造，既保留了古朴自然的历史感，又注入了现代的时尚元素。设计不仅继承了老厂房的工业气质，还通过创意布局和设计使其变成了时尚的现代办公空间。

对设计师来说，空间的解读和经历百年岁月的木构架是绝美之处。设计尊重原有的空间、原有的材质和原有的历史风貌，节制地使用设计语言，保护文物本体，同时达到精确的设计效果。纯白色和绿灰相间的 BOLON 地胶板和百年的木构架一起展现了整个空间的性格。必需的照明和空调也隐在白色柜体和木构架之后。强烈的色彩冲击和新旧对比创造了一个传统的超现代空间。

传统的纺织车间为了防止丝织品被强光照射，都是北光高窗。而作为办公空间，这样的采光和照明均不理想，这是设计的最大难点。经过仔细推敲，设计保留了北光窗、增加了新风系统，并将原有的木质结构整体涂成了白色，使空间更为明亮。照明则采用二次照明，更柔和、更接近自然光，也让办公空间更为舒适。

2014 年 6 月，京杭大运河申遗成功，本案作为杭州世界文化申遗点段改造老工业厂房的典型项目，以其鲜明的特色和独特的魅力赢得了来自各界的肯定。犹如一颗明珠，其与其他同运河密切相关的物质文化遗产一起构成了运河丰富的人文景观。

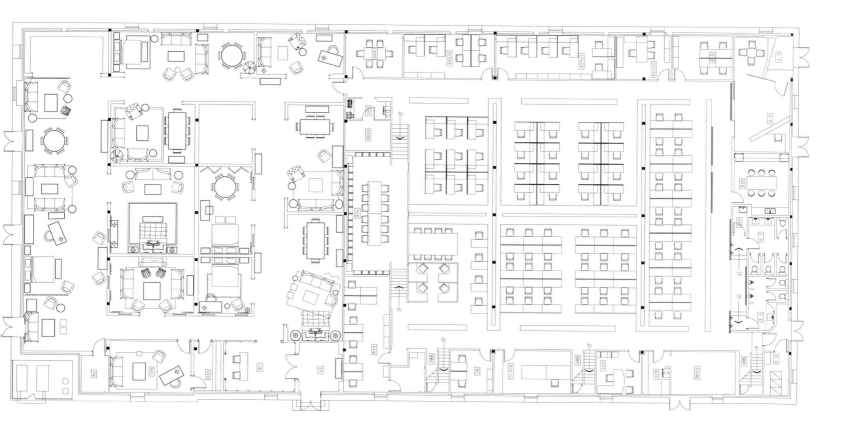

Walmart Brazil

沃尔玛巴西办公室

DESIGN COMPANY
Estudio Guto Requena
ARCHITECTURE, ENGINEERING, LIGHTING DESIGN, INSTALLATION
Lumini
COLLABORATOR
Tatiana Sakurai (Architect)
DEVELOPMENT, CO-DESIGNER
Paulo De Camargo (Architect), Lucas Ciciliato (Architect)
LANDSCAPE
Camila Opipari
RENDERING
Lucas Corato
LAMP DESIGN
Guto Requena, Vitor Reis, Lucas Ciciliato
INTERN
Lucas Miilher
CONSTRUCTION AND CONSTRUCTION DOCUMENT
Uficcio
WOODWORK
SPJ Móveis
FURNITURE
Tidelli, Tok Stok Dpot, Maurício Arruda, Meu Móvel De Madeira, Fernando Jaeger, Depósito Kariri, Bortolini, Flexform
VISUAL COMMUNICATION
LEN Office
CARPETING
Escinter, Citycase
LOCATION
Brazil
AREA
6,400 m²
PHOTOGRAPHER
Fran Parente

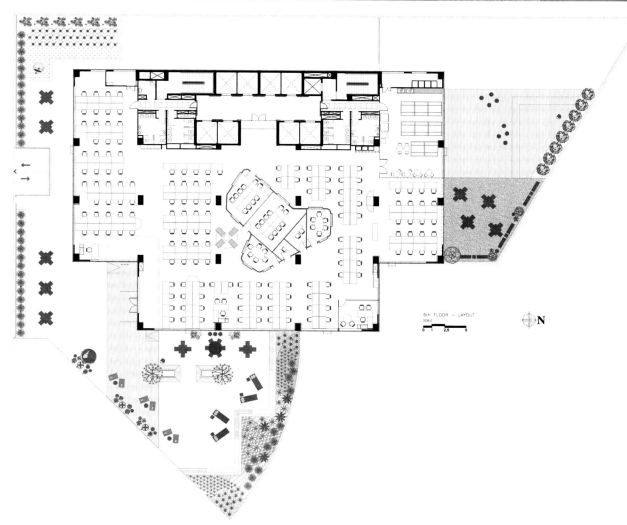

6th FLOOR — LAYOUT

The office design was derived from a research methodology developed by Estudio Guto Requena. Interviews and dynamic online exchanges with company employees were conducted to assess values, needs and expectations. Three principal focal points emerged from this process: digital culture, the "Walmart.com" brand and "brasilidade" (Brazilian identity). This research also informed the choice of colors, materials, forms, programming and design concepts.

The designers applied these three focal points and their commonalities to an exploration of the building's prominent terrace and developed a guiding concept for the company's headquarters: the Urban Veranda. Design choices referenced the Brazilian habit of engaging outdoor areas for social interaction and relaxation. Elements included beach chairs, large buildings with terraced facades, picnics (visible in the carpet patterning), the patios and balconies of Brazilian homes, and the rural habit of placing a chair in the street to enjoy the evening and chat with neighbors.

The headquarters occupied six floors with over 1,000 m^2 each. One of the challenges of this project was to bring a more human dimension to the work environment with spaces that were welcoming and comfortable, even pleasant and informal, while maintaining professionalism and practicality.

The designers prioritized the use of domestic furniture in both the offices and lounges, with signed pieces by the established Brazilians designers Maurício Arruda, Jader Almeida, Lina Bo Bardi, Paulo Alves and Fernando Jaeger. The designers also included pieces that were part of the popular Brazilian imagination, such as rocking chairs, beach chairs, porch chairs and picnic tables. For the production of objects and decorative elements they used images of contemporary Brazilian photographers, as well as maps, illustrations and Brazilian folk art. Skateboards and bikes referenced the lifestyles of younger employees.

The outdoor area was designed for both work and relaxation. Wood decking ordered the environment, together with porch furniture, shaded areas, a space for yoga and a grandstand facing the facade that could host small events, concerts and film screenings. A mini-golf course was also specially designed for the terrace.

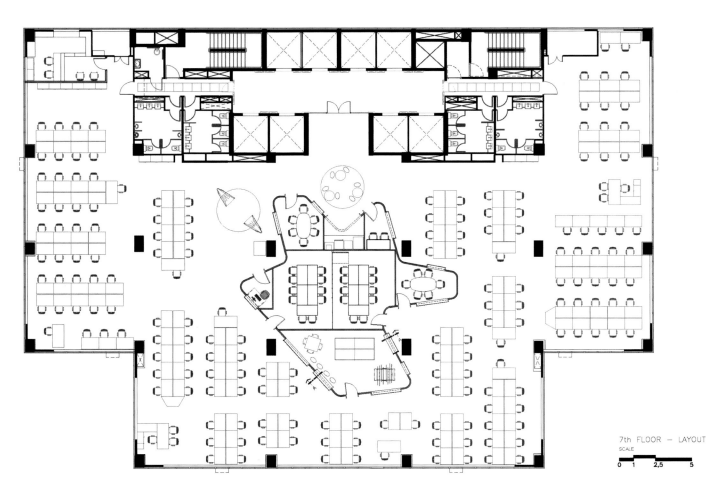

7th FLOOR — LAYOUT

8th FLOOR - LAYOUT

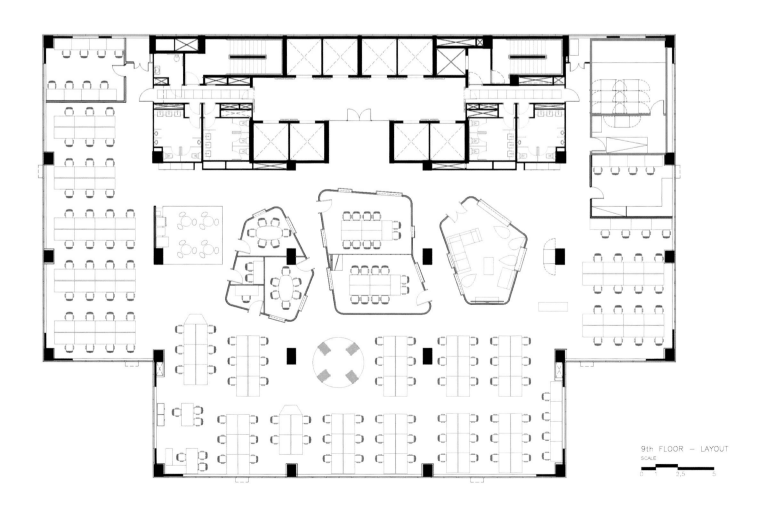

9th FLOOR - LAYOUT

办公室设计源于设计单位 Estudio Guto Requena 的研究方案。其通过访谈和与公司员工的动态在线交流来评估价值、需求以及期望。在此过程中，显现出三大主要焦点问题：数字文化、"沃尔玛.com"品牌以及"巴西特色"。这项研究也涉及到色彩、材料、形式、策划以及设计理念方面的选择。

设计师们运用了这三点并融合其共性来打造此建筑的阳台，并以此来确立公司总部的设计理念——"都市阳台"。设计选择考虑到了巴西人喜爱户外社交和休闲的习惯。设计师在空间设计时引入的元素包括沙滩椅、带有阳台式立面的大楼、野餐元素（体现在地毯图案中）、中庭、巴西家庭式阳台以及那种沿袭乡村习俗在街上放一把椅子以供人们享受夜景，和左右邻居聊聊天。

总部一共有六层，每层大约 1000 m²。其中，该项目的一大挑战是在工作环境中融入更多的人性尺度，缔造出一种友好、舒适甚至愉悦的非正式空间，同时还要顾及专业性和实用性。

办公室和休息室两处，设计师们优先考虑使用巴西产家具，另外也有公认的巴西设计师毛里西奥·阿鲁达、玉·阿尔梅达、丽娜·柏·巴蒂、保罗·阿尔维斯以及费尔南多·耶格等人设计的产品。此外，他们还采用部分代表巴西特色的家具，包括摇椅、沙滩椅、门廊椅和野餐桌。当代巴西摄影师的照片、地图、插画以及巴西民间艺术也贯穿于物件生产和装饰元素当中。滑板和自行车反映了年轻员工的生活方式。

户外空间设计集工作和休闲于一体。木质天台确定了环境的风格，还设置了门廊家具、阴凉区、瑜伽室和一个面对外立面的看台区，在这里可以举办小型会晤、音乐会和电影放映等活动。阳台上还特别设计了一个迷你高尔夫球场。

Deliqatê Restaurant

Deliqatê 餐厅

DESIGNER
Fernando Forte, Lourenço Gimenes, Rodrigo Marcondes Ferraz
DESIGN COMPANY
FGMF Arquitetos
COORDINATOR
Ana Paula Barbosa, Marilia Caetano, Sonia Gouveia
EMPLOYEE
Carolina Matsumoto, Juliana Fernandes, Raquel Engelsman
INTERN
Felipe Bueno, Gabriel Mota, Gabriela Eberhardt, Patrícia Kupper, Rodrigo de Moura
LOCATION
São Paulo, Brazil
AREA
275 m²
PHOTOGRAPHER
Rafaela Netto

Occupying a lot at Alameda Jaú in the Jardins district of São Paulo, Deliqatê is a restaurant specialized in gourmet sandwiches and is the first enterprise from this group of investors whose intention is to take it to other capital cities in the country.

The 5 m×30 m terrain was previously occupied by a two-storey building from the 1940s. After being closed for about fifteen years, it was totally deteriorated and was demolished in order to give space to the new building. The architects preserved the lateral and back walls, which leaned against the neighbors and were built with solid terracotta bricks. They were carefully preserved and reinforced during the brickwork so that the brick-surface aspect became part of the project after it was finished, as a kind of reminiscence of what was there before the new building.

The new building may be understood as a big steel grey structure that is introduced in this small space between the neighbors and the street. This structure aims to support the slabs of the three storeys and the roofing, also locking the neighboring walls.

Due to the lot's high unevenness, the restaurant has three storeys— the ground floor, where the entrance, front desk, counter, table area and outdoor deck and terrace are; the basement, where the kitchen and other service spaces are; and a mezzanine-shaped upper floor that stoops over the entrance and has indoor and outdoor table area, and may be cut off from the ground floor in case of closed events. The roof slab is occupied by a water tank and air-conditioning, exhaustion and air inlet systems, etc.

The architects seek a permanent contact between the building and the street. So they do not use fences or walls, and the 5 m mandatory setback is occupied by a deck surrounded by plants, as well as a front totally filled with glass. The 6 m glass are divided in two sections: the lower one has sliding panels, which allow the entrance of clients and the communication between the internal area and the deck.

FRONT VIEW

SECTION AA

1 PRODUCTS EXHIBITION
2 DINING ROOM
3 W.C.
4 EXTERNAL LOUNGE

FIRST FLOOR PLAN

1 MAIN ENTRANCE
2 DECK
3 WAITING ROOM
4 HOST \ CASHIER
5 DINING ROOM
6 ACESSIBLE TOILET
7 EXTERNAL LOUNGE
8 ACCESS TO UNDERGROUND

GROUNDFLOOR PLAN

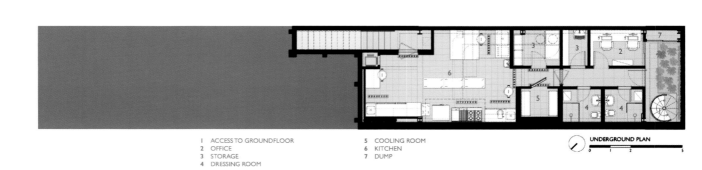

1 ACCESS TO GROUNDFLOOR
2 OFFICE
3 STORAGE
4 DRESSING ROOM
5 COOLING ROOM
6 KITCHEN
7 DUMP

UNDERGROUND PLAN

Deliqatê 餐厅位于圣保罗雅尔丹区 Alameda Jaú，是一家专门做美味三明治的餐厅，它也是该区域第一个投资者想要在其他首都城市开分店的公司。

20 世纪 40 年代，这一 5 m×30 m 的地带曾经为一个两层的建筑所有。在封闭了 15 年之后，原有建筑完全损坏，为了给新的建筑物腾出空间，其最后被拆除。建筑师们保存了倚靠在隔壁建筑的用陶砖堆砌而成的侧墙和后墙。它们被小心地保存下来，在砌砖过程中还被加固，所以砖块表面在建筑完工后也成为这项工程的一部分，这让人们回忆起新建筑之前的旧建筑是什么样子。

这座新建筑可能会被看成一个在街道与相邻建筑之间狭小的空间中的大型灰色钢架结构。这个钢架主要是为了支撑这三层楼和屋顶，而且也锁定了相邻建筑的墙壁。

由于建筑用地极度的不平整，饭店被分为三层，一层为餐厅入口、前台、柜台、就餐区和户外露台；地下一层为厨房和其他服务空间；阁楼状的上层楼面弯向入口处，里面有室内和室外的就餐区，如果有封闭的活动时可与一层隔离开来。屋顶板部分有水箱、空调、排气装置和进气装置。

建筑师们希望能够建立起建筑与街道的永久联系，所以他们没有使用围栏或是墙，而是建造了 5 m 长的被植物环绕的平台和玻璃临街面。6 m 高的玻璃被分为两个部分。下半部的玻璃有滑动功能，顾客可以通过它进到餐厅里面来，也使内部和露天平台之间的交流更加方便。

Emporium Santa Isabel

圣伊萨贝尔购物中心

DESIGNER
David Guerra, Laura Rabe
DESIGN COMPANY
David Guerra Architecture and Interior Design
PROJECT TEAM
Gisela Lobato, Gabriel de Souza, Guilherme França, Jefferson Gurgel, Laís Machado, Nínive Resende
LOCATION
Belo Horizonte, Minas Gerais, Brazil
AREA
1,050 m²
PHOTOGRAPHER
Jomar Bragança

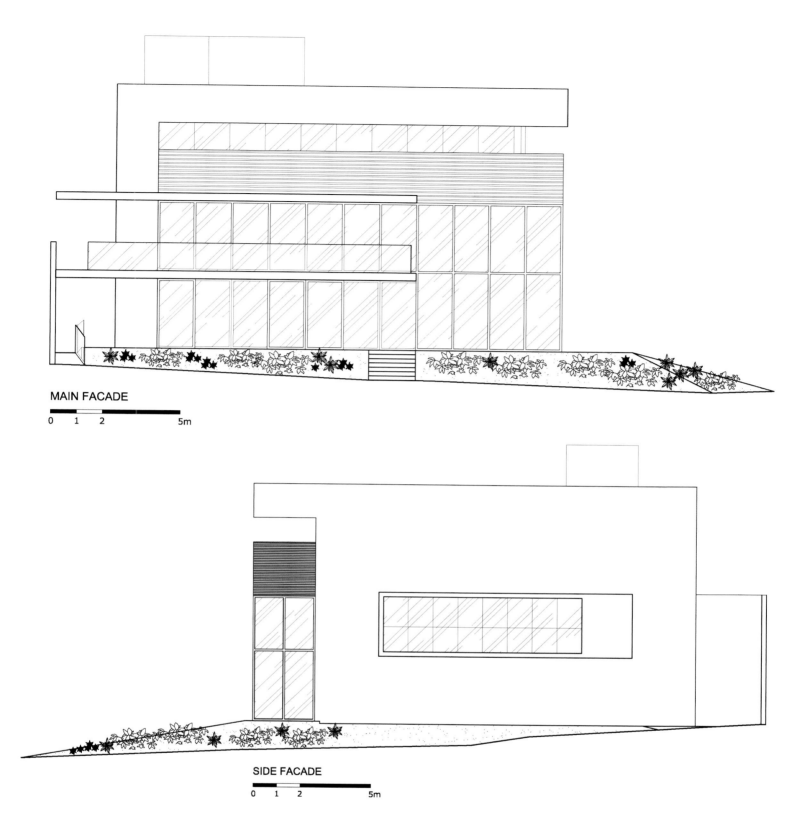

MAIN FACADE

SIDE FACADE

The idea for Emporium Santa Isabel is to build a technological and contemporary architecture in a more affective way, referring to the tradition of the countryside and baroque style of Minas Gerais. The resulting building displays a metallic structure coated with automotive paint, clay textured concrete panels, glass panels with aluminium golden colored frames, a balcony with a bamboo ceiling, walls covered with old metal fences and plant vases, gardens with fruiting trees, and bordeaux, golden, copper and colonial blue as theme colors of the brand identity. These appealing textures, colors and volumes, combined with the privileged location of the building on a corner, help local people to develop a special affection for the place, even helping to care for the gardens.

In the basement, there are the storage, dressing room, and bread production area.

On the ground level, the bakery, patisserie, emporium and auxiliary kitchen are located. The products are displayed on wood and metal shelves, with a light and fluid design, which provide easy visualisation of the products.

On the first floor, with a view to the garden, there is a buffet, a pizzeria and a special products market. A footbridge links this space with the outside balcony, with a bamboo ceiling. Bamboo is also used for a sculptural lamp that goes through the whole side of the emporium, creating a cosy atmosphere and an impact illumination. There are also an auxiliary kitchen, a administration office and restrooms.

On the second floor, there are the patisserie production, storage room, staff resting area and restrooms.

Sustainability is present in many factors. For instance, the metallic structure, pre-fabricated ceiling, and dry-wall panels make construction fast and reduced waste production. The LED illumination system reduces energy consumption, and natural illumination reduces the need to use artificial lights during the day. Cross ventilation replaces artificial air conditioning. Local market and materials are given priority, such as the hand-made bamboo items, bought from local artisans, and the wood and clay textured concrete, which is developed by the architects' office, being applied for the first time in Brazil on a facade.

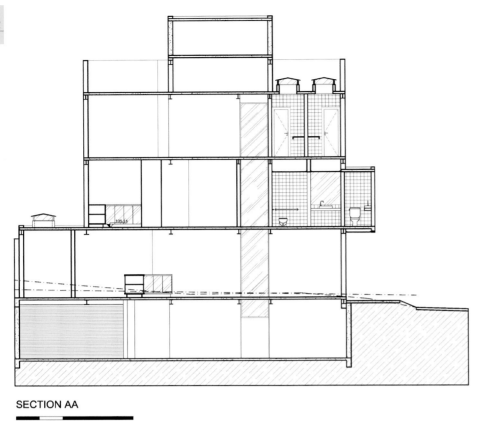

SECTION AA
0 1 2   5m

圣伊萨贝尔购物中心的空间设计理念是以更具情感的方式建设一个结合技术和现代的建筑大楼，也就是以充满传统的乡村风格和米纳斯·吉拉斯的巴洛克建筑风格的方式。已建成的大楼以涂有汽车漆的金属架构的建筑元素为主，黏土纹理的混凝土墙板、镶有金色的铝制外框的玻璃嵌板、带有竹制天花板的阳台、被覆盖着旧金属栅栏和植物花瓶的墙体，还有正果实累累的花园以及以波尔多红、金色、铜色和蓝色为主的主题颜色，都是这家商场的品牌标识。这些优美的建筑纹理、色彩和外观以及商场独特的角落位置都使当地人对它产生了特殊感情，有的人甚至会帮忙打理花园。

地下室设有仓库、更衣室和面包生产区。

一层是面包店、蛋糕店、商场和辅助厨房。产品摆放在木质或者金属展架上，流畅迷人的灯光设计使顾客更容易直观地看到他们购买的产品。

二层面向花园方向，可以看见自助餐厅、披萨饼店和特殊产品市场。一条人行桥将这个空间和带有竹质天花板的阳台连接起来。为了营造舒适惬意的氛围和灯光效果，用竹子做成造型灯也安装在商场的一侧。这层楼也设有一个辅助式厨房，还有行政办公室和洗手间。

三层设有法式糕点生产区、储藏室、员工休息室和洗手间。

这个建筑可持续的设计理念体现在方方面面。例如，采用金属架构、预制天花板和墙面板，不仅加快了建设速度还减少了废物产生。LED照明系统减少了能量消耗，自然光照明降低了在白天使用人工照明的需要，对流通风的设计也代替了人工空调。建筑师们也优先选择在当地购买材料，例如，商场内的竹子工艺品就来自本地手工艺者。还有第一次在巴西使用在外立面的由建筑设计师所在公司开发的木质和黏土纹理的混凝土板。

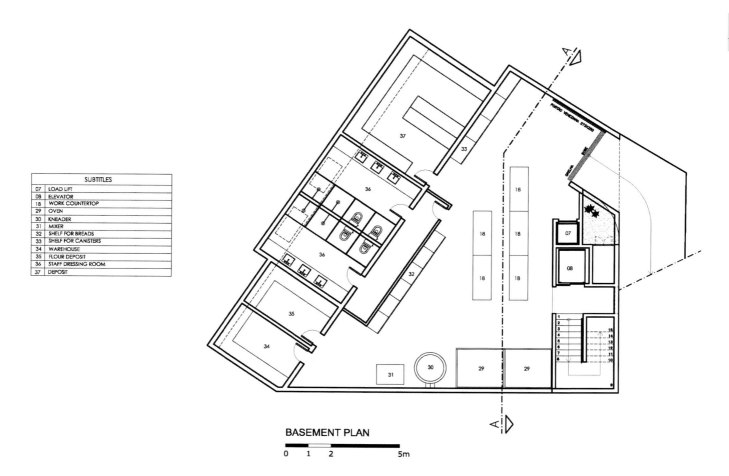

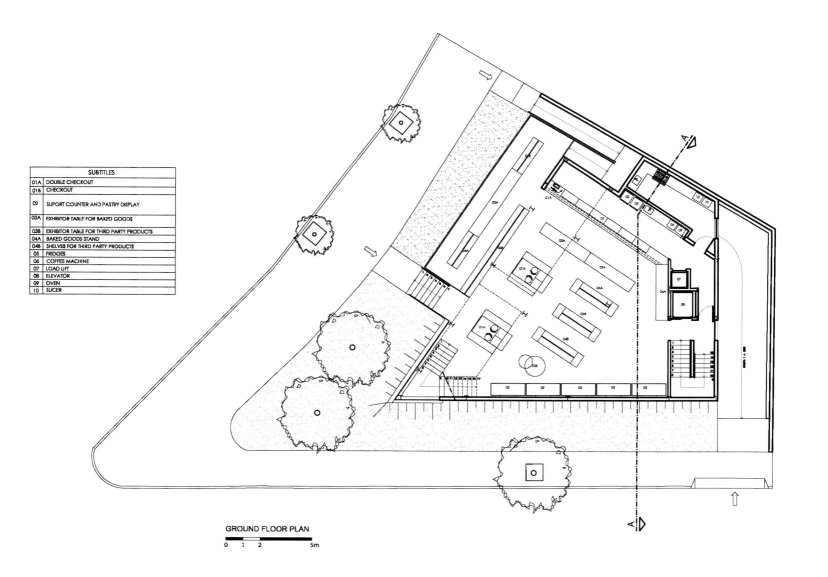

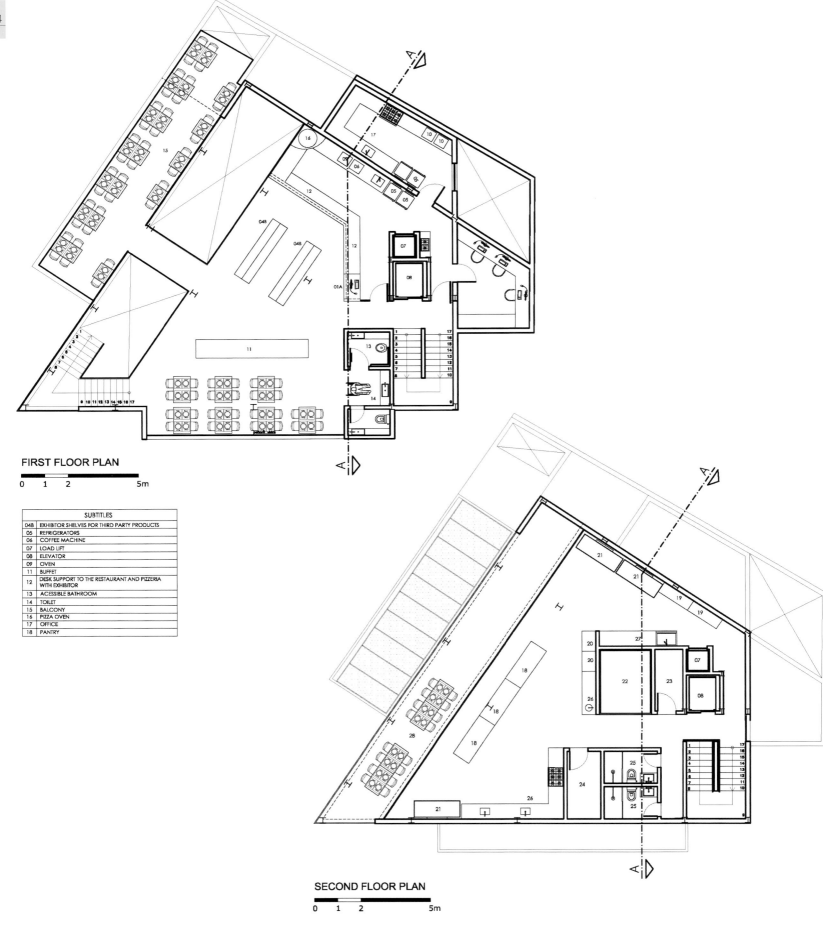

265

Hunt a Lobster Restaurant

猎寻龙虾餐厅

DESIGNER
Yuriy Yumashev, Yuriy Cherebedov, Dmitriy Zhukov
DESIGN COMPANY
Seventh Studio
LOCATION
Moscow, Russia
AREA
683 m²
PHOTOGRAPHER
Alexandr Usanov

Have you ever hunted a lobster? Now you can do it. Choose one from aquarium and it will go to open kitchen. You can see how it is cooked.

The best seafoods that fished near Russian seashores get to the restaurant in 24 hours after they are hooked. So it makes Raw Bar possible. Menu is written by hand. Food are so fresh that there is no time to reprint menu.You can eat a great choice of raw dishes made from seafood.

Customers can relax on soft sofas between brutal metal elements, vintage lamps and green plants, and look at Moscow center through the panoramic windows and choose pan Asian food.

The restaurant has 2 floors. Bar and several seatings are on ground floor. Show kitchen, aquariums and robata are on the 1st floor. During planing the designers found out that they needed to make new bigger stairs and to take away a piece of slab making connection between floors more comfortable.

Floor is finished by natural wood and tiling. Wall elements are made from aged wood, moss, gypsum aged panels, reinforced glass, old bricks and metal.

Most of the elements are made specially for this restaurant. Gypsum aged panels are designed by the design company in 3D. It is made by cnc machine from MDF and then silicone form for gypsum is made. After fixing gypsum panels, decorator colors them in vintage style. Some elements are bought but changed by decorator.

Now you can see all the ventilation elements decorated by metal net. Wood stove for show kitchen is so big that the designers have to take away some facade glass to get it in by crane.

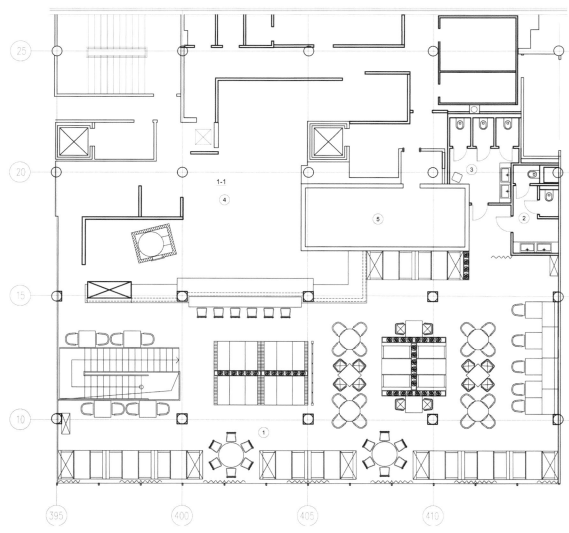

你曾猎寻过龙虾吗？现在你可以做到这一点。从水族箱里选择一只龙虾，它会被带到开放式厨房，在那里你可以亲眼目睹龙虾是如何被烹饪的。

从附近的俄罗斯海岸捕捞到的最好的海鲜，在24小时之内就可以到达餐厅。因此使海鲜餐厅的菜品品质得到保证。菜单是手写的，因为食物太过新鲜来而不及重新打印。你可以品尝到各种由海鲜烹制的生鲜菜肴。

客人可以坐在绿植环绕的的软沙发上放松心情，四周布满冷酷的金属元素和复古灯，还可以一边欣赏全景窗外的莫斯科街景，一边享用亚洲美食。

这家餐厅有两层。餐吧和一些座位在一层，开放式厨房、水族箱和烧烤炉在二层。设计师们设计餐厅的时候认为应该建造新的更宽敞的台阶并撤走一块混凝土石板来其使楼层之间的衔接更舒适自然。

地板是由天然木材和瓷砖拼接而成。墙壁装饰材料有古老的木头、绿色的青苔、石膏面板、强化玻璃、古旧的砖块以及金属。

大部分的装饰元素都是为这家餐厅特制的。例如，陈旧的石膏板是设计公司通过三维设计而成，并用CNC数控操控机器切割中密度纤板，然后制造出生产这个石膏板的硅胶模型。固定好这些石膏板后，装潢师将它们打造成复古的风格。还有一些购买的饰品也经过了装潢师的修改。

所有你可以看到的通风设备都用金属网做了装饰。由于开放式厨房中的木制火炉太大，设计师们不得不撤掉一部分玻璃幕墙才能用起重机将它放置进去。

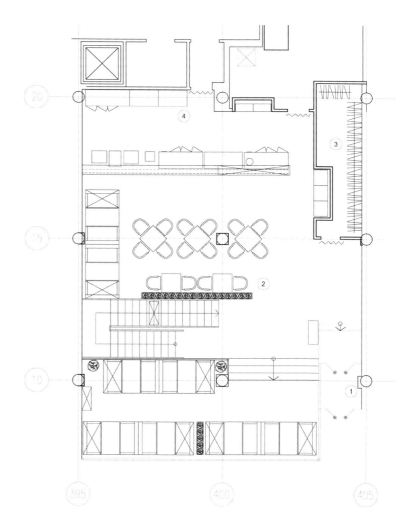

Il Giorno Canteen DM

白日食堂设计手册

DESIGNER
Olga Treivas, Vera Odyn,
Anastasia Gomberg, Vlada Rodionova
DESIGN COMPANY
FORM Bureau
CLIENT
Semifreddo Group
LOCATION
Moscow, Russia
AREA
610 m²
PHOTOGRAPHER
Asya Baranova

The new Il Giorno canteen is located in a 19th century weaving factory – the Danilovskaja Manufacture.

The space is divided into three main parts: the bar and canteen close to the entrance, the main dining space and the kitchen. High vaulted ceilings and original brickwork harmoniously blend with new elements, such as partitions and furniture from perforated metal and poured concrete flooring. The spatious main dining area is characterized by rows of columns that are clad with tiles of one color but different shapes, thus bringing together the entire room.

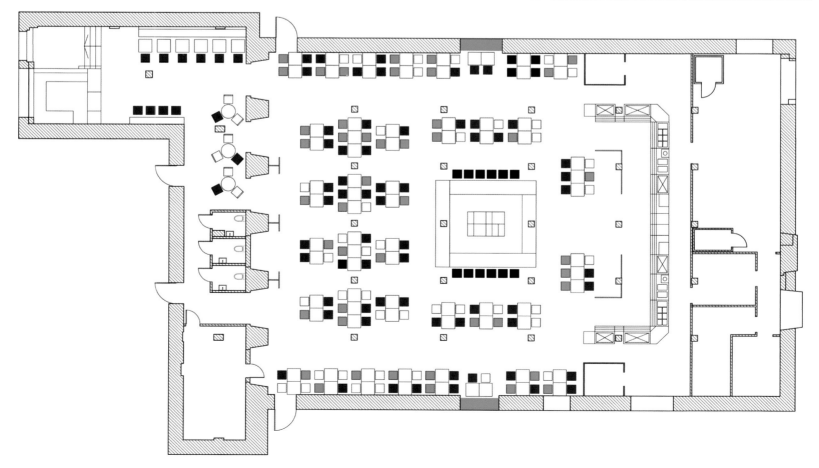

新的白日食堂坐落于一间 19 世纪纺织工厂——Danilovskaja 制造厂。

空间主要分为三部分：位于入口处的吧台和餐台、主用餐区和厨房。高大的拱形天花板与新颖的砖墙同新元素完美融合。这些新元素包括钢板制成的隔断和家具，以及混凝土灌浇的地板。用餐区域最大的特色是一排排覆盖着瓷砖的立柱。瓷砖虽颜色统一但形态各异，使整个房间成为一体。

Korean Dessert Cafe Mu-A

韩国 Mu-A 甜品咖啡馆

DESIGNER Won Myoung Hyun	
DESIGN COMPANY INEX Design	**LOCATION** Korea
BUILDER Design Chang	**AREA** 139.5 m²
CLIENT Geum Sik Yoon	**PHOTOGRAPHER** Jae Yoon Kim

In a forest of heavy concrete buildings, turning towards inside of the downtown main road to narrow and dense alleys, located near Yakjeongolmok of Donseongro, center of Daegu, the Korean style dessert cafe Mu-A is a space for time travel with fragrant tea in an alley where modern and contemporary history of Daegu is alive.

The punched louver showing exterior scenery also plays a role of bringing the strong afternoon sunlight and surrounding scenery into the interior space more softly by filtering them.

The louver is derived from the traditional Bunhapmun (a sliding door to shut the plank-floor room off from the yard), and to which color of Hanji (traditional paper made from a mulberry tree) is applied.

It expresses the layer of traditional space with skip floor by borrowing the method from a village with level difference by natural slope, and presents dynamic human traffic line and attractions.

"Welcoming" space on the stereobate of the first floor tells the start of space.

Human traffic line becomes complicated due to the interior space with six steps, which influences service style of each space.

Centering around the counter table on the first floor, there are a main kitchen and an auxiliary kitchen on the basement and the second floor each.

Meals are sent from the main kitchen to each floor through a dumbwaiter, and served in each floor. To employees, human traffic line of multi-storey building is more complicated than that of one-storey building.

However, the human traffic lines from the hall of the second floor to room and service area are just half floor up, giving diversity to the space through the repetition of wideness and narrowness.

Ceiling of the third floor is high in order to give openness to visitors on the top floor, to attempt natural movement through human traffic line, and to arrange the "tea room" – the most sacred place in this building.

Wooden ceiling frames woven in three dimensions emphasize the identity of the building and play a role to stabilize the hanging tea room visually.

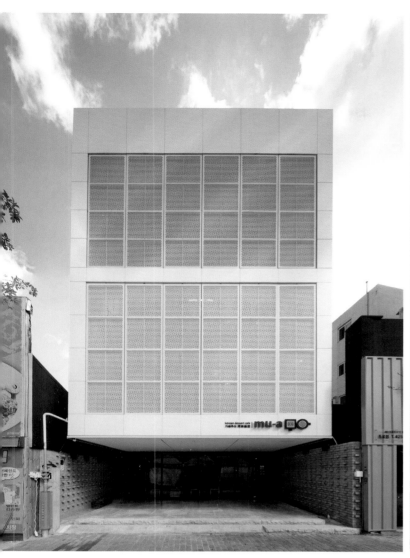

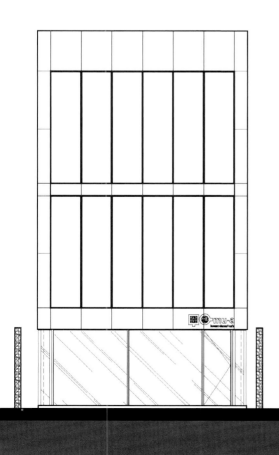
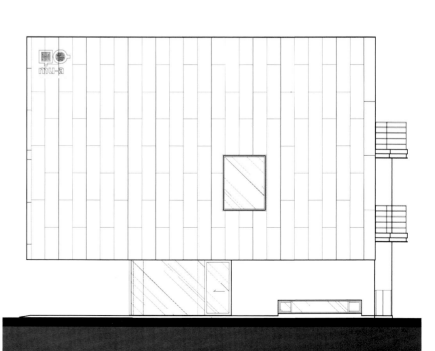

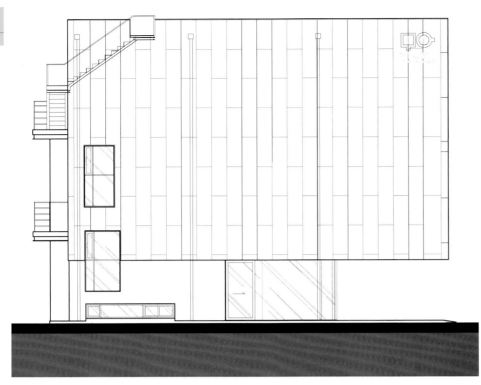
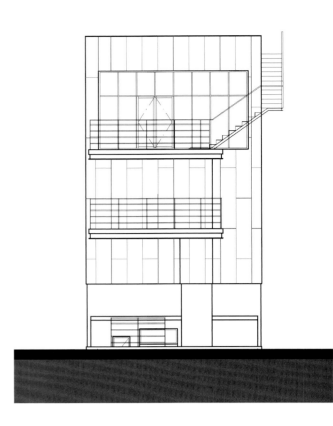
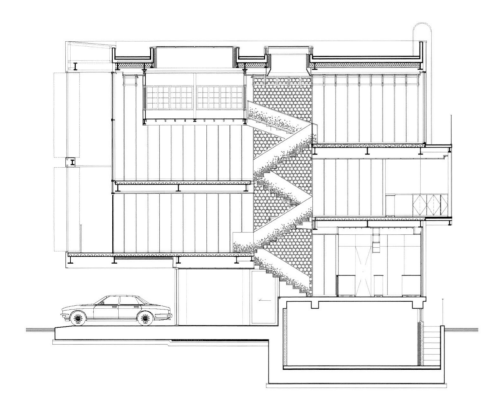
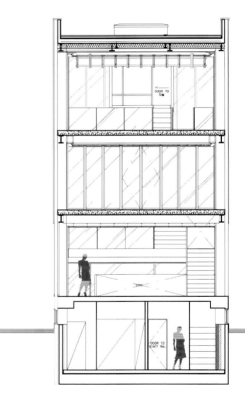

在一片混凝土建筑群中，走进市区主干道来到拥挤狭窄的小巷，Mu-A 韩式甜品咖啡馆坐落于大邱市中心 Donseongro 的 Yakjeongolmok 附近。这是个在小巷中享受香茶使人仿佛进入时空旅行般的空间，在这里大邱市的现代和当代历史都鲜活了起来。

穿孔的百叶窗不仅让人们欣赏到外面的风景，还使强烈的午后阳光和周围的景色透进屋子时更柔和迷人。

百叶窗的设计灵感来自传统木门 Bunhapmun（一种可以将木板房与院子阻隔起来的滑门），并选择了褚皮纸（褚皮制成的传统纸）的颜色做修饰。

它表现了跃层房屋传统空间的层次感，通过借用从一个村庄得来的方法，利用自然斜坡产生的水平差来设计，从而展示了动态的室内交通线和建筑风格。

"迎宾"空间设置在一楼屋基，代表着整个空间的开始。

由于室内空间设计了六个台阶，这影响了每个空间的服务风格，使整个室内交通线变得更加复杂。

柜台设在一楼，以其为中心，地下室和二楼都配有主厨房和辅助厨房。

菜肴的准备大多是在主厨房中进行，再通过餐用升降机送至各层。对于员工来说，多层建筑的室内交通线要比单层建筑复杂得多。

但是，从二楼大厅进入房间和服务区走一半楼梯就到了，这段室内交通线让人们感受到狭窄空间和开阔空间的交织重叠，使设计空间显得更加多元多样。

为了使客人在顶层感到开阔宽敞，能够在室内交通线上自然地活动，设计师们在三层楼设计了很高的天花板，并设计了一个茶室，创造出整个建筑内最梦幻的空间。

木材交织而成的立体天花板框架彰显了整个建筑的风格，还在视觉上起到了稳定悬空茶室的效果。

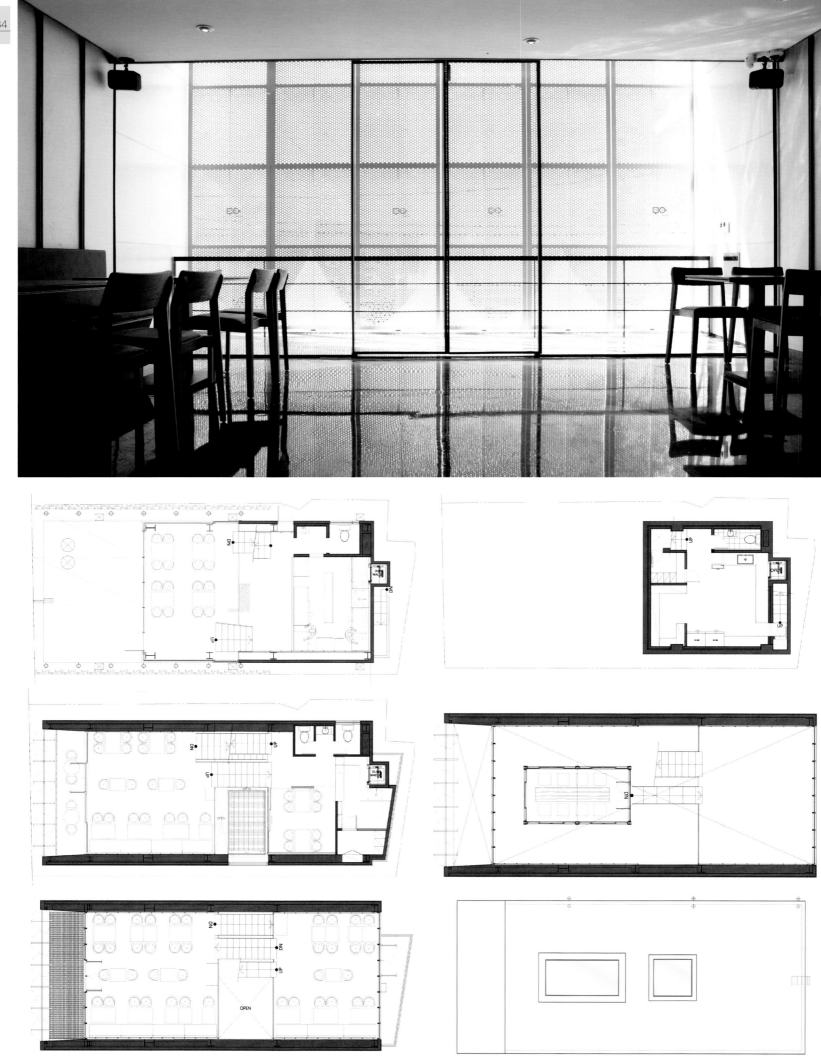

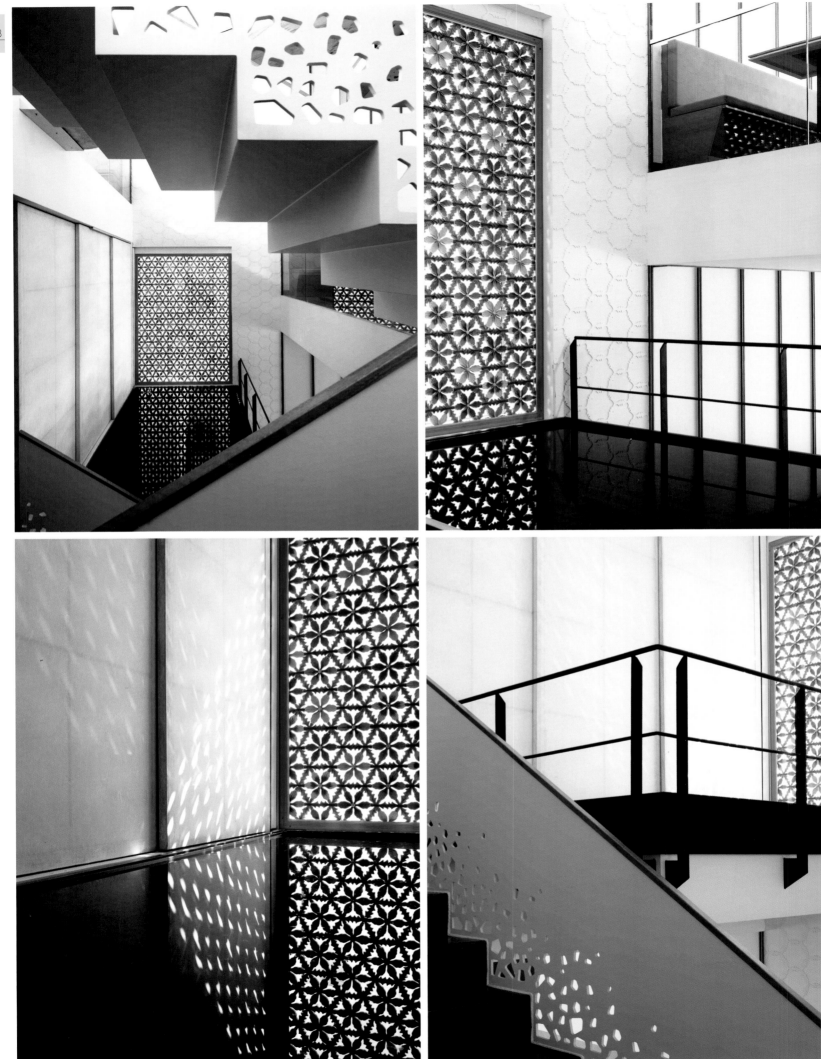

289

Murakami Restaurant

村上日式餐厅

DESIGNER
Yuriy Yumashev, Yuriy Cherebedov, Dmitriy Zhukov
DESIGN COMPANY
Seventh Studio
LOCATION
London, UK
AREA
582 m²
PHOTOGRAPHER
Craig Howart

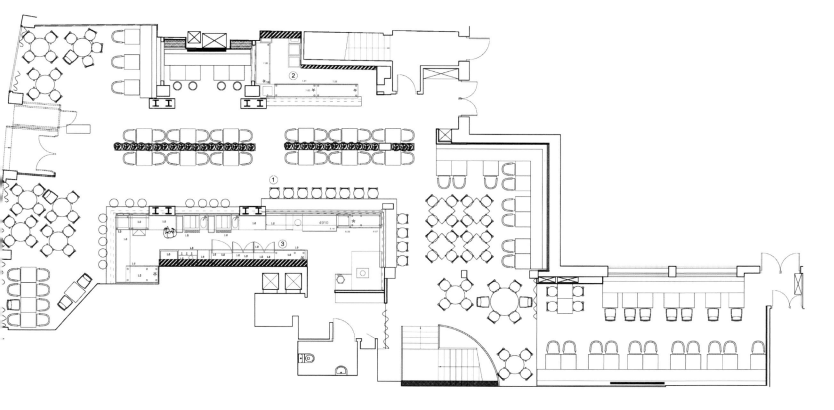

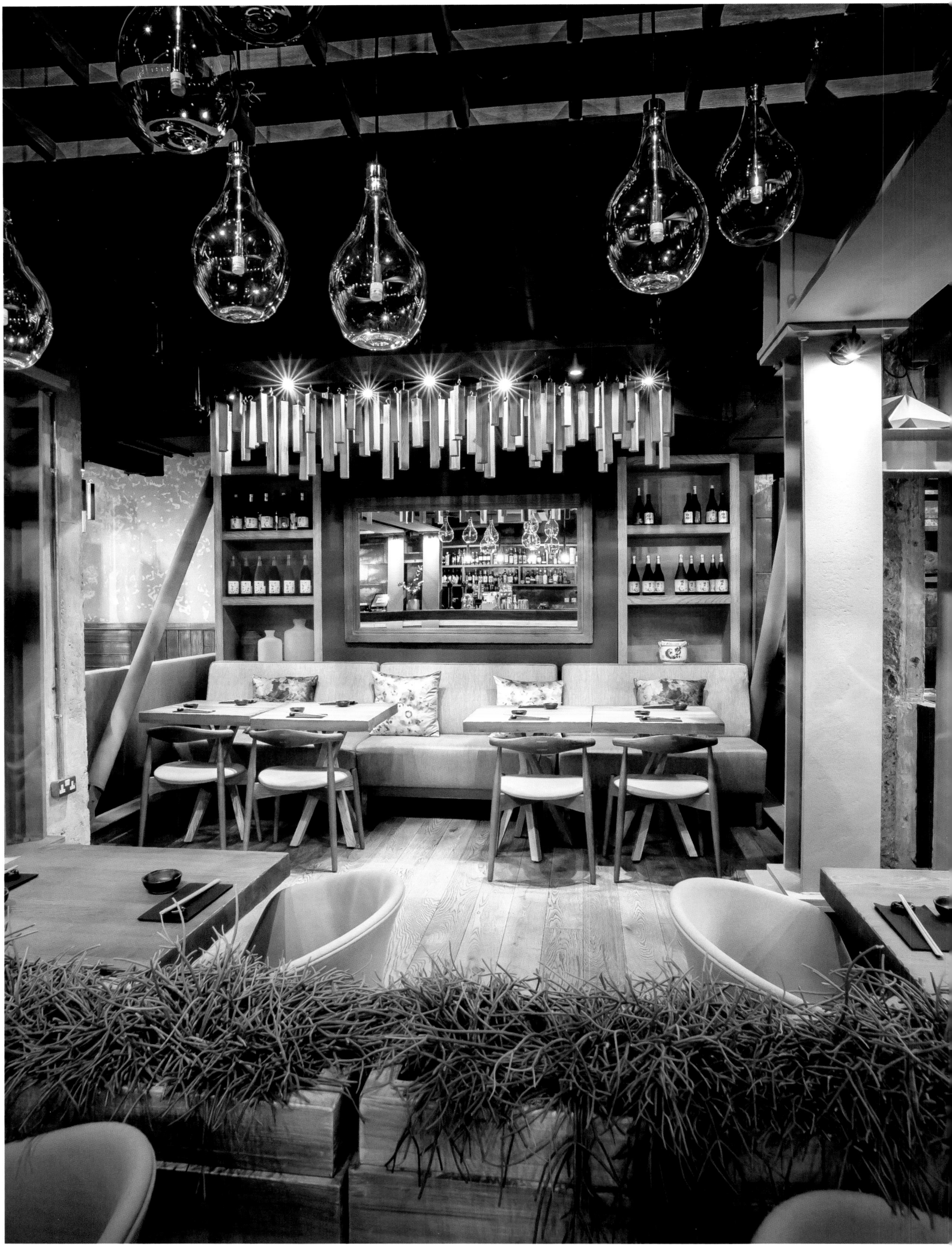

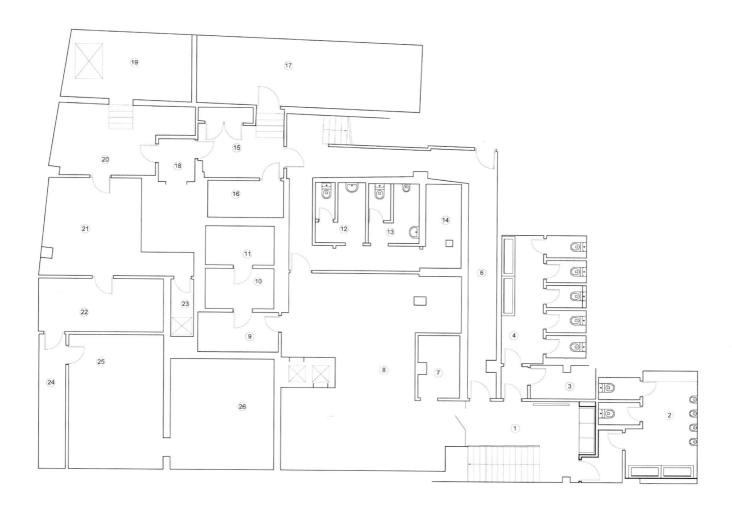

The project is a contemporary Japanese dining on St Martin's Lane in Theaterland, London. The restaurant is based on 2 floors. Bar, show kitchen, robata, sushi bar and seating are on the ground floor. Kitchen, offices and WC are on the underground floor. Dishes and dirty plates go to and from the kitchen by two different elevators.

Design is a compilation of loft and modern Japanese elements. Most of the walls and columns are cleared from gypsum covering to show their original state. The designers find elements that are made during years of using this premises in different purposes. So there are concrete, different types of bricks (small and big), old stucco and metal structural elements. All structural elements are colored by fire resisting paint. Some walls are decorated by wooden covering to make interior more warm. One covering looks like squama, another looks like fortification. One wall is covered by moss to add more freshness to the restaurant.

Floors are made from natural wood and hexagonal tiling. Bars are made from aged wood.

The designers leave open HVAC systems but hide electrical wires. Wooden frames of different types are made on the ceiling to decorate it. They are arranged to make lighting look more nice and to hide some places of ceiling. Some chandeliers are made in Ukraine and Italy, and some are bought in UK. On one brick wall the designers use "garage" lamps as lighting. They are specially ordered for this restaurant. Windows are changed into panoramic windows. Stair railing is left as it is before, handrail and steps are renewed.

Food preparing is made as show processes. You can see sushi rolling, robata grilling and cocktail making. Grill is closed by glass to defend customers from heat.

The design company is situated in Ukraine, so the designers have to fly to London twice a month to direct building process.

　　村上日式餐厅是伦敦戏剧街圣马丁巷的一家现代日本料理餐厅。这家餐厅共有两层。酒吧台、开放式厨房、烧烤炉、寿司吧和座位在一层。厨房、办公室和卫生间在地下一层。菜肴和用过的餐具分别通过两个不同的电梯运送出入厨房。

　　村上日式餐厅的设计融合了loft高大而敞开的设计风格和现代日本风情元素。设计清理了屋内大多数的墙壁和柱子上的石膏表面以展现其最原始的风貌。设计师们认为这些元素的形成来源于多年来这个场地被用于不同的用途，因此就会有不同类型的混凝土砖块（或大或小）、古老的石灰泥和金属架构等建筑元素。所有的建筑元素都涂上了防火漆。为了使室内更温馨，一些墙壁上还用木质材料做了装饰。有的墙壁装饰得像鱼的鳞片，有的则像堡垒。还有一面墙上铺满青苔，给整个餐厅增添了清新的气息。地板是由天然木材和六角形瓷砖拼接而成。酒吧则是由旧的木头制成。

　　设计师们安装空调系统的时候注意隐藏了电源线。天花板用不同类型的木质框架装饰。它们使得室内的灯光看起来更迷人并遮住了部分棚顶。天花板上的吊灯产自乌克兰或者意大利，还有的是在英国购买的。有一面砖墙上设计师们使用了安全灯，这些灯是为餐厅特殊定制的。餐厅的窗户全改为全景式窗。楼梯栏杆没有变，但是翻新了扶手和台阶。

　　在这里，食物的制作过程成为一场表演秀，你可以亲眼目睹寿司卷、烧烤和鸡尾酒的制作过程。但是为了防止顾客烫伤，烧烤架用玻璃封闭了起来。

　　由于公司位于乌克兰，设计师们不得不每个月两次飞往伦敦来指导装修。

Factory 5

Factory 5 自行车商店

DESIGN COMPANY
LINEHOUSE
LOCATION
Shanghai, China
AREA
150 m²
PHOTOGRAPHER
Benoit Florencon

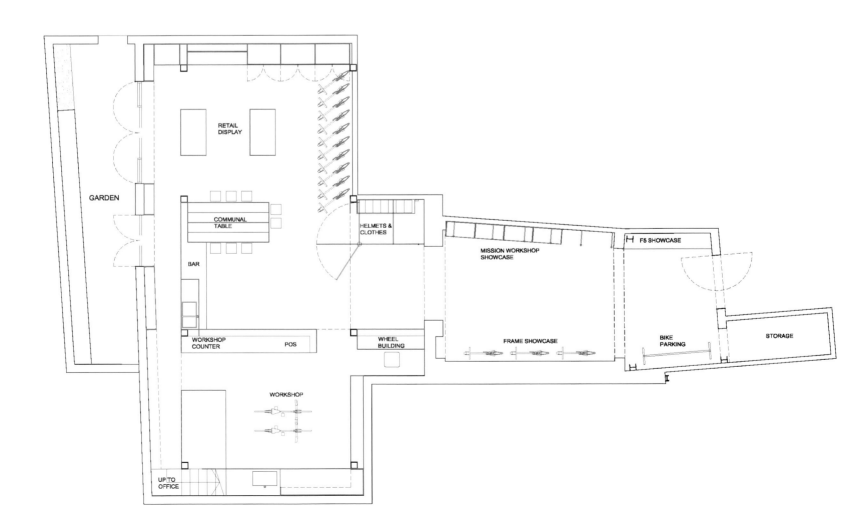

The new Factory 5 shop nestled to the rear of a former bustling local marketplace on Changhua Road in the north of Shanghai's Jing'an District. The bicycle shop occupied a double height space within the existing walls of the previous market made up both interior and exterior spaces. The walls were lined with grey Chinese brick and rough concrete walls and ceilings. These surfaces were retained as the rough canvas for the shop. LINEHOUSE inserted a mezzanine to create office space on the upper floor and located the bicycle workshop below. The design of the mezzanine and the workshop counter used 150 mm × 150 mm I-Beams welded horizontally between the black metal columns. The I-Beams wrapped around the space for the cashier counter as well as a bar. The mezzanine was wrapped with blackboard painted MDF for a high display surface for wheels and bicycle frames.

The first floor retail and workshop space focused on retaining a cycle hub feel with the creation of a bar serving coffee and beer and a wooden plank communal table. The raw concrete floor of the market was overlaid with a clear epoxy and concrete walls fold up to become the walls of the counters.

Feature retail display walls were arranged to highlight Factory 5's signature bicycle frames, Mission Workshop bags and clothing, and component and product displays. Material sourcing was as local and salvaged as possible: reclaimed wood from Shanghai's demolition sites was used for the table tops and the cabinetry, gas pipe installations and salvaged army trunks form the shelving and racks for clothes and bags. A small garden space to the rear was planted with bamboo and lined with a long bench.

新的 Factory 5 自行车商店位于上海市静安区北部昌化路上，这里曾经是一个当地的生产市场。自行车商店拥有双倍层高的空间，先前市场留下的墙壁同时围合出内部和外部空间。这家商店的墙体铺排着灰色的中式瓷砖并设计有粗糙的混凝土墙面和天花板。得以保存的旧墙面做为背景，就像是早已存在的商店中的画布一样。LINEHOUSE 工作室设计了一个夹层，来创造一个上层的办公空间，把自行车展示放在下层。

夹层和展示柜台的设计使用了 150 mm × 150 mm 的工字钢，钢材水平焊接在黑色金属柱之间。工字钢环绕于这个空间形成收银柜台和吧台。夹层用漆成黑色的中密度纤维板覆盖，使其成为一个自行车轮和车架的高空展示表面。

一层设计的零售空间和展示空间重点在于保持一种自行车轮毂的感觉，吧台可以提供咖啡和啤酒，还有木质共享桌。商店销售空间的水泥地面涂着透明环氧树脂，水泥墙面则成为柜台展示墙。

特色零售展示墙的设计是为了突出 Factory 5 的标志性的自行车架、Mission Workshop 的背包和衣服及其他相关零件和产品。建筑材料尽量选择采购当地的和废弃的物品，例如，用上海拆迁场地回收的木材制作桌面和橱柜，用燃气管道设施和废旧军用卡车做成衣架和背包架。建筑后面的小花园空间种植了竹子，并且摆放着一个长椅。

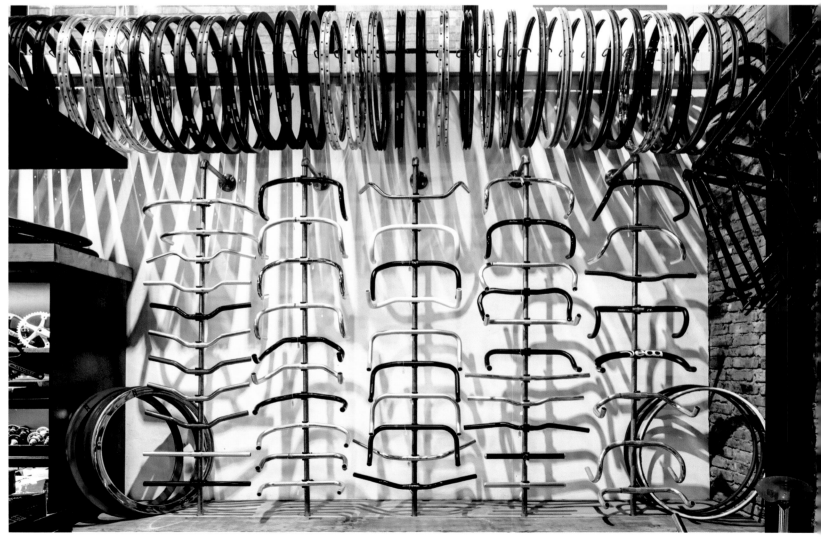

Russian Performance

俄罗斯车库展览会

DESIGNER
Olga Treivas, Vera Odyn, Asya Baranova, Inna Vostrikova
DESIGN COMPANY
FORM Bureau
CURATOR
Yulia Aksenova, Sasha Obyhova

CLIENT
Garage Museum of Contemporary Art
LOCATION
Moscow, Russia
AREA
1,100 m²
PHOTOGRAPHER
Yuri Palmin, Ilya Ivanov

The Garage Museum exhibition "Russian Performance: A Cartography of its History" explored a century of the medium's history and its unique traditions in Russia.

The exhibition was structured chronologically, with the peculiarities of each decade or epoch reflected in the architectural design of the corresponding section of the show. The exhibition space was doubled by constructing an additional level, through which visitors traveled along a carefully curated route.

The first hall of the exhibition covered the decade from 1910 to 1920 and presented the artistic experiments of the Russian avant garde. Visitors entered a large space with the grand set design for the Magnanimous Cuckold – a futurist opera. The exhibition stood in the space inspired by avant garde compositions. The choice of materials, including wood, black and red metal, also referenced the period.

After a long pause spanning several decades, the performance genre returned to Russian art in the 1970s. Ascending a panoramic ramp, visitors entered a viewing platform with a "modernist" interior in green and birch plywood.

Stepping into the 1980s, performance art continued to address a small circle of people, who delved into the problematics and meaning of closeted events, which were held in the artists' private home or studio, or "independent public spaces" such as parks and fields.

During perestroika in the late 1980s, the situation changed dramatically. Socio political changes throughout the country gave artists a chance to change the format of their work, to gain a wider audience, and to employ new technologies. The exhibition space tried to reflect it with a more exuberant palette, such as metal foil and pastel pink.

The 1990s was a period of growing activism, a new generation of movements, which were characterized by "direct action" and the search for new conflicting forms of interaction with the public. For this exhibition space, untreated grey floor boards and unfinished partitions were chosen to reflect the atmosphere of the "reckless 1990s".

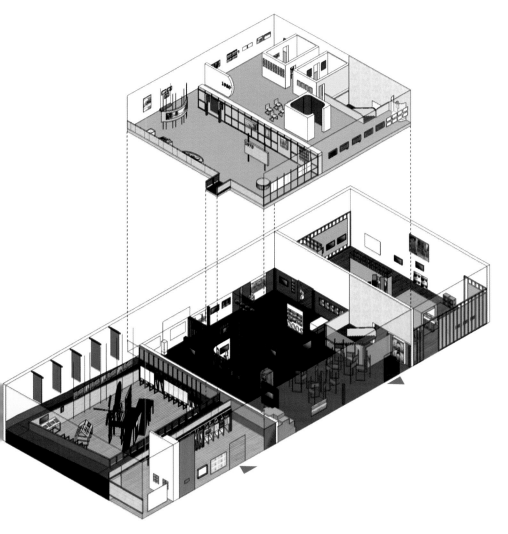

车库博物馆"俄罗斯表现：历史再现"展览发掘了100年来俄罗斯的中期历史和独特的传统。

本次展会按时间顺序安排，每一个十年或时代的特别之处会体现在展览相应部分的建筑设计上。该展览空间多搭建一层，面积变成之前的两倍，参观者可以沿着这条认真规划好的线路进行参观。

展览的第一展厅介绍了1910年至1920年，并展示了当时前卫的艺术实验。参观者进入了为《绿帽王》——一部未来主义歌剧设计的整体空间。位于这个空间的展览主要受前卫的创作作品所启发。在原料的选择上，有木头、黑色金属和红铜，也反映出相应的时代。

在几十年的漫长暂停后，展览的表现类别又回到了20世纪70年代的苏联艺术。通过全景上升坡道，参观者进入了一个观景平台，这是用绿色桦木胶合板装饰的具有现代特色的内部。

步入20世纪80年代，行为艺术继续在一小部分人中展开，这些人致力于钻研各种问题和探讨在艺术家的私人家庭或工作室，或"独立公共空间"，例如，公园和空地举行私下密谈的意义。

20世纪80年代后期重组改革期间，情况发生了巨大的变化。在全国各地社会政治的变化带给艺术家一个改变他们的工作形式，获得更广泛的受众，并采用新的技术的机会。为了反映这段历史，该展览空间试图运用更加明朗的色调，例如，金属箔片和水粉红色。

20世纪90年代是行动主义的时代，新一代的动作特点是"直接行动"，同时要寻找新的冲突形式与公众互动。这部分展览空间中，未处理的灰色地板和未完成的分区被选来反映"鲁莽90年代"的气氛。

306

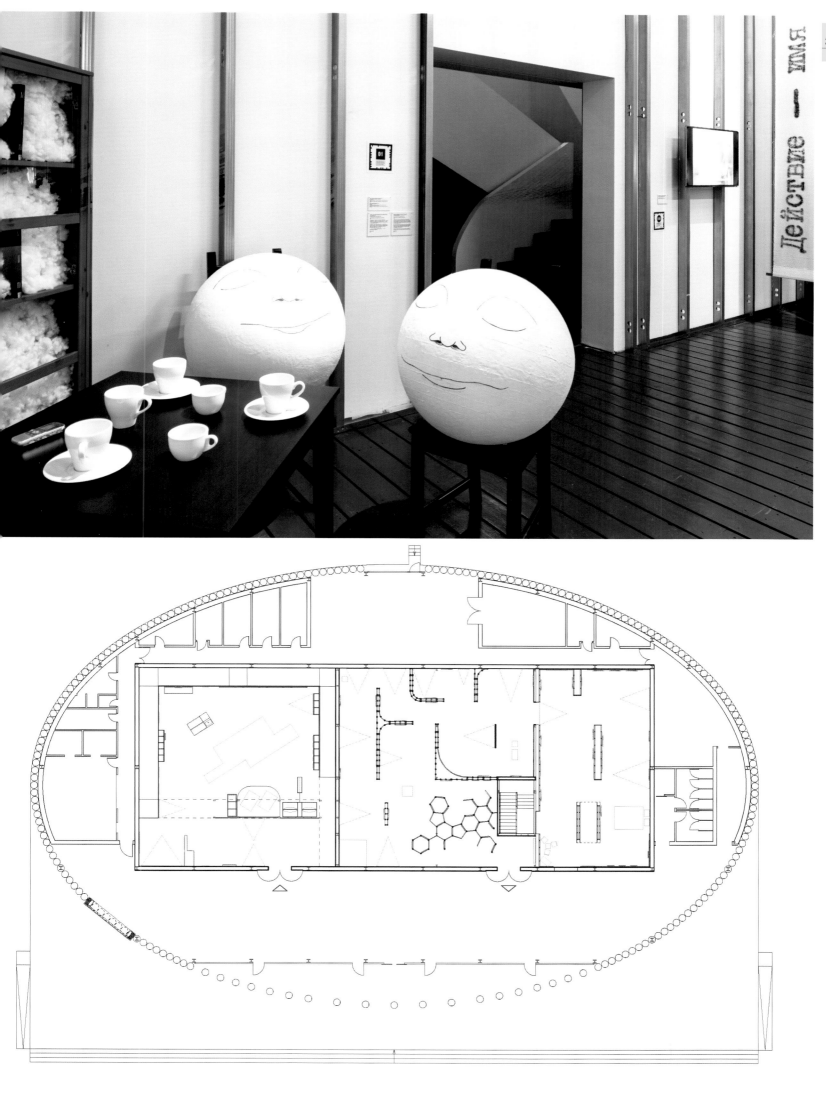

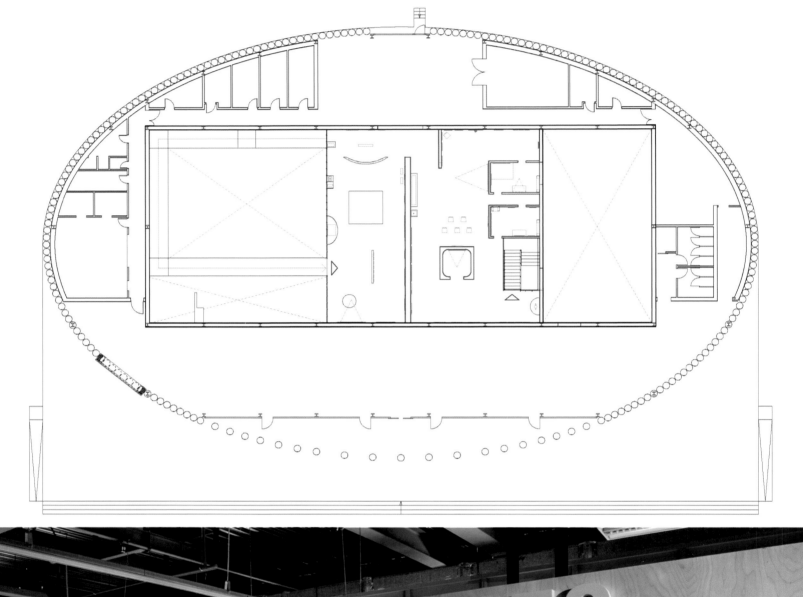

INDEX

索引

FGMF Arquitetos

Created in 1999, FGMF produces contemporary architecture, without any restraints regarding the use of material and building techniques, seeking to explore the connection between architecture and its environment.

In these few years of existence, we've had the opportunity to deal with a wide range of architectural scales, what enhances our belief that, just as life itself, architecture ought to be plural, heterogeneous and dynamic. Urban realm, cultural facilities, residences, sports complexes, hospitals, schools, commercial buildings and many others are part of the same urban landscape and of our daily life: knowing how to deal with all these programs is a way to enrich our design, in contrast to a specialized architecture.

Based on the professional and academic experience of its associates, FGMF has an innovative and inventive approach. There are no pre-conceived formulae: at every challenge we start from scratch, using design as a research tool.

Our dedication and hard work led us to the satisfaction of receiving relevant national and international awards, among which some from the Instituto de Arquitetos do Brasil (IAB), Living Steel, Editora Abril and Dedalo Minosse. In 2010, we were the most awarded practice in Brazil.

Recently, FGMF has been chosen as the only Brazilian office to integrate the Architects Directory from Wallpaper Magazine. Our practice was also selected as one of the Emerging Architects of 2010 from Architectural Record Magazine. Our work has been published in more than 15 countries and took part in both national and international exhibitions. More than simply recognition of a well done job, it works as motivation for creative projects.

blitz

Design Blitz

Design Blitz was founded by Partners Melissa Wallin and Seth Hanley in 2009. As a full service architecture and interior design firm, they provide the complete range of architectural services required to take a project from programing through construction. Though their focus is the built environment, they are committed to total design solutions – balancing buildings, branding, and experiences.

Design Blitz leaves many of the inefficient conventions of traditional practice behind by systematizing delivery, removing unnecessary management layers and leveraging technology to reduce errors, paper, and shorten the feedback loop wherever possible. By removing the barriers to open communication and flattening the delivery process, they deliver high quality projects faster.

In short: They Blitz It.

D+DS Architecture Office

D+DS Architecture Office was founded in 2007 by Ellen Depoorter and Jeroen De Schrijver in New York.

The office has a holistic approach to architecture and design. The experience of the user, the connection to the local context and the impact on the environment are at the basis of each project. The design is generated through the vetting and playing out of different concepts which feed the design. D+DS Architecture Office works in the spirit of enquiry, challenging preconceptions and testing conventions in collaboration with the client and consultants.

The concept informs the structure, shape, materials, the ecology and energy performance of the building and the relationship of the building to the urban environment. This holistic approach is augmented by a strong commitment to the clients D+DS Architecture Office serves and to the public domain and the many users involved. A high degree of personal service, coupled with respect for the precious resources of cost and time, characterise our client relationships.

D+DS Architecture Office combined 35 years of experience ranges over a wide variety of scale, geographic location and program. The projects D+DS Architecture Office works on range from urban projects to office towers, hospitality and residential projects, retail and work environments to art installations.

UNK project

UNK project is the Russian architectural bureau with the western principles of work. The company was founded in 2000 by Yuliy Borisov, Julia Tryaskina and Nikolay Milovidov. All three architects have the experience of working abroad. Since 2011 the team also has worked in partnership with leading British practice Scott Brownrigg. The result of cooperation of the Russian and British architectural schools is the development of unique high-tech objects and the town-planning projects corresponding both the recent western trends and features of the domestic market of real estate.

UNK project specializes in architectural and engineering design of buildings and complexes, design of corporate interiors and public spaces. Among the realized and current projects of architectural bureau is reconstruction of the pool "Luzhniki", residential quarters in innovative park "Sokolovo", front faces of the "Metropolis" mall, housing estate "Dutch Quarter", administration and storage complex the Bork, offices of the companies Mail.ru, WALT DISNEY STUDIOS, Google, Microsoft, Samsung; the interiors of "Russian Song" theater, the retail chain "The Gourmet Globe" and "The ABC of Taste", Bocconcino restaurants, theater chain "Luxor" and so forth.

The projects of UNK project repeatedly received prestigious professional awards, among which are European Property Awards, CRE Awards, Urban Awards, Best Building Awards, etc. The international partner of UNK project is the company with centenary history, which has been one of top ten architectural bureaus of Great Britain for several years. Architects of Scott Brownrigg workshop have unique experience in the public and industrial sector, hotel business, the transport sphere, town planning.

LINEHOUSE

LINEHOUSE was established in Shanghai in 2013. The studio seeks projects of varying scales and typologies that allow them to explore both the poetic idea and pragmatic solution. Each project is an unexpected response to the client, physical context, brief and budget created through a process of dialogue, investigation, drawing, cutting, and collaboration.

LINEHOUSE has international experience in design and construction from the small scale and the intricate to the grand and public. They are drawing on their experience to create new ideas and outputs emphasising qualities of construction, materials and light.

DRAFT Inc.

DRAFT Inc. is a Tokyo-based interior design firm founded by Taiju Yamashita in 2008. Our designers specialize in the interior design of office, as well as restaurant and commercial space. We aim to create "cycle of happiness" by our design with everyone who visits the space because we believe good design delivers greater creativity and productivity to human and organization. No matter what type of projects, workspace and community, big and small, classic and modern, our focus is always on "how to design more effective space where foster creativity and promote corporate purpose, value, and vision with limited space and budget". In 2013, DRAFT opened two new offices in Shanghai (China) and Cebu (Philippines). We have been serving the community by embracing the mixed idea of local culture and latest trends from Tokyo.

Maurizio Lai

Born in Padua in 1965, Maurizio Lai studied at the Faculty of Architecture in Venice and then at the Milan Polytechnic, where he graduated.

He started his career early in 1998 as set designer, in collaboration with RAI Mediaset, the Italian national TV channels, while designing shop and display concepts for LONGINES, MONT BLANC and Gucci, as well.

Later on, he would join a group of international planners to develop major hospitality projects with Viaggi del Ventaglio and THE CLUB MED.

In 1998, he founded LAISTUDIO in Milan, building up synergies between architecture and design which reflect in the creation of different impacting concepts for homes, shops, showrooms, spas, restaurants and resorts.

In 2005, he won the international competition to design the W HOTEL MIAMI for Starwood Hotels & Resorts Worldwide.

Since 2008, Maurizio Lai holds a strategic alliance with the South African group DHK.

Next to the professional practice, Maurizio Lai cultivates an interest in education and promotion of the Culture of Design.

An original interpreter of contemporary aesthetics, his work finds impulse in the instinct, while exploring the concept of creation, in a quest for knowledge, where matter is the master of it.

Quadrangle Architects

Quadrangle Architects can be summed up in two words: inspiration realized. The firm's creative vision as architects is always grounded in practical realities. Since founding the firm in 1986, they have consciously built a diverse and wide-ranging portfolio. The firm has never focused exclusively on a particular sector. Instead, they focus on a particular kind of client: visionary and idea-driven, yet anchored by pragmatism. Every project Quadrangle undertakes is guided by a set of beliefs that, appropriately, has four sides. First, they believe that architecture is about solving real-world problems; the spaces they create are shaped by business insights. At the same time, each project must be part of a larger strategy; what they really build are relationships (and many go back decades). Third, the firm believes in thoughtful design that solves complex challenges while respecting budgets and deadlines. And lastly, everyone in the firm is passionately committed to innovative thinking and advancing the evolution of design.

Haworth Tompkins

Haworth Tompkins was formed in 1991 by architects Graham Haworth and Steve Tompkins. Haworth Tompkins' London-based studio has designed buildings in the UK and elsewhere for clients across the public, private and subsidised sectors including schools, galleries, theaters, concert halls, housing, offices, shops and factories.

The buildings of Haworth Tompkins are influenced by the specific chemistry of individual places and cultural situations. Haworth Tompkins put enormous effort into understanding a building's context and the needs of its users, a process which often yields original or unconventional solutions. What they have in common is an approach rather than a stylistic signature.

Professionalism and sustainability in the widest sense underpins our creative output. The combination of a relatively small, close-knit team and a carefully limited number of projects in development at any one time enables them to offer an exceptional level of service to clients. Haworth Tompkins' work has won over 50 design awards and is published internationally.

IWAMOTOSCOTT ARCHITECTURE

IwamotoScott Architecture

IwamotoScott Architecture is an award winning architecture and design firm established by Lisa Iwamoto and Craig Scott in 2000. Based in San Francisco, California, the firm has gained national and international recognition for innovative design with projects around the country and overseas. Their client list includes arts organizations, educational institutions, media firms, commercial developers, and private clients, and projects consist of work at all scales including urban design, buildings, interiors, full-scale fabrications, museum installations and exhibitions, and theoretical proposals. Each project is treated as a unique opportunity to achieve outstanding design quality regardless of project scope or scale.

IwamotoScott Architecture is at the forefront of employing architectural technology and computation. Their design approach proceeds from in-depth research coupled with creative experimentation. Responses to project circumstances and client desires are tested through extensive iterative modeling, drawing, and rendering using both physical and digital media. Whenever possible, ideas are also tested with full-scale mock-ups to evaluate their interactive, tactile and experiential potential. They enjoy working closely with clients, collaborators, consultants and contractors to realize the best possible architectural solutions.

Ector Hoogstad Architecten

Ector Hoogstad Architecten is an architectural office in Rotterdam with over 60 professionals; an inspiring mix of experience and young talent. The firm is led by Joost Ector (design principal) and Max Pape (managing director). We believe that well-designed buildings are extremely valuable to the quality of our everyday lives. They make us feel and perform better, individually as well as collectively. To us architecture is happiness. We excel at the design of working environments, research facilities and public buildings for government, education and culture. Transformation is gaining importance in our portfolio. Our design solutions are simple, sustainable and user friendly. Our open source working method, based on user participation and integral design, is unique.

Sergey Makhno Architect

Sergey Makhno is an architect, designer, as well as workshop project manager. In 2003 he founded Makhno Workshop. Then it was a desire of one man, and now Sergey is working together with a big talented team.

"Being a leader always requires taking huge responsibility and having aspiration to reach lofty goals. I love what I do, and with my team we can implement design projects of any complexity. I'm always in progress, striving forward, because I want my clients to have special homes, restaurants and offices...", Sergei Makhno says.

The best ways of spending his holidays are visiting international design exhibitions and plunging into antique shops. Sergey may come up with an idea of not only well-forgotten and obsolete things but also with brand new ones.

IND Architects Studio

Amir Idiatulin is CEO of IND Architects Studio. IND Architects Studio is featured by a particular attention to details. The firm believes that it is details that show the quality of architecture.

IND Architects Studio consists of experienced architects who develop the projects starting with a sketch and following it up to complete implementation of intended ideas.

IND Architects Studio is united by true passion and commitment.

Since the studio was founded in 2008, they have been dealing with design of apartment and public buildings, town houses and interiors, office premises, hotels, business-centers and restaurants. In these latter days, we are engaged in development of landscape and urban design areas and widely participate in various competitions.

The studio strengths: individual attention to every client, effective system of business processes, flexible and quick decision-making, complex approach to architecture and interior, thorough researches and quality analysis at all project phases, quick adaptability to new market requirements.

IND Architects Studio provides clients with comprehensive design documents prepared in accordance to high standards.

Their clients are comprised of those who generate the demand for high-quality architecture: individuals, state officials, businessmen, and development corporations.

You can find their works in the most picturesque places of Russia, Spain, Montenegro and Kazakhstan. IND Architects Studio is keen to expand this list and take part in up-market international contests.

Looking for substantial cooperation!

za bor architects

za bor architects is a Moscow-based architectural office founded in 2003 by Arseniy Borisenko and Peter Zaytsev. The workshop's objects are created mainly in contemporary aesthetics. What distinguishes them is an abundance of architectural methods used both in the architecture and interior design, as well as a complex dynamical shape which is a hallmark of za bor architects projects. Interiors demonstrate this feature especially brightly, since for all their objects the architects create built-in and free standing furniture themselves. Many conceptual and realized design-projects by za bor architects were awarded at international exhibitions and competitions. At the moment za bor architects is involved in variety of projects in several countries. za bor architects have been involved in more than 60 projects including residential houses, a business center, a cottage settlement, and many offices. Among the clients of za bor architects there are IT, media and government companies such as Badoo, Castrol, Forward Media Group, Yandex, Inter RAO UES, Iponweb, Moscow Chamber of Commerce and Industry and others.

David Guerra Architecture and Interior Design

Founded in 2002 and located in Belo Horizonte, Brazil, the office David Guerra Architecture and Interior Design develops projects in the areas of residential, commercial and institutional architecture, interior design and furniture design.

With more than 23 years of activity in the field, architect David Guerra combines creativity and elegance to develop unique spaces full of soul, memories, concept and identity.

The office's works have been presented at the 14th Biennale di Venezia 2014, in the exhibition "Time Space Existence" organized by the G.A.A. Foundation, among other exhibitions. They have also been extensively published in Brazilian and international press, and received several awards such as the Silver A' Design Award for Interior Space and Exhibition Design; IAB Architecture Award, Casa Claudia Interior Design Award, Amide Interior Design Award and Olga Krell Interior Design Award.

GBD

GBD Architects Incorporated

Ask anyone at GBD Architects Incorporated about the kind of buildings we make and they're likely to tell you the same thing: GBD makes buildings that work hard. Just like we do. They'll say that these buildings should be honest and wise, and give out more than they consume. They should be thoughtfully and artfully crafted, always placing utility before vanity. They should be quiet, majestic solutions to real human problems, not the source of new ones. They will tell you that building is a privilege, not an excuse for public theater, not a showcase for the latest technology or flavor-of-the month material, not for awards or notoriety.

This drive for a better building, for a decidedly more human building, is rooted in the firm's history. GBD was established in 1969 in Portland, Oregon, by a cadre of senior architects who left the city's largest and most prestigious architecture firm to create something that didn't exist: a practice that would excel in three primary areas. GBD would do a better job of understanding the people they were working with and what their business objectives were, so they would meet and exceed those objectives; GBD would work harder to consider how each space or building would be used and enjoyed, so they be aware, from the very beginning, that a legacy was inevitable. That every decision and interaction would have an impact in some way: the kinds of buildings they built, how they built them, and the way they interacted with each other, their clients and the community.

The experiment has enjoyed many successes. In terms of meeting the needs and goals of clients, many of those early GBD clients continue to be some of our most active clients today, as they themselves have prospered and expanded.

INABA

INABA

Jeffrey Inaba is the Principal of INABA.

INABA specializes in an analytical approach to design and construction. The office applies a creative method of problem solving to arrive at unique options that can be realized with a high degree of resolution. This stems from the practice's broader philosophy to grasp the depths of a problem and to provide clients with findings that assist in making better-informed decisions during concepting, programming, design, and realization phases. INABA's analysis-based method is the foundation for its brand and content work. The services of INABA include concept development, content creation, and media design. INABA works on improving cities, brands, and the experience of environments.

Soesthetic Group

Soesthetic Group is an exclusive design boutique with a prime focus on architecture and interior design for public and residential buildings. Soesthetic Group works with major Ukrainian businesses, foreign entities and individuals.

The company was founded by Nataliya Shchyra and Vickoria Oskilko in Ukraine in 2008 and has offices in Kiev and London, England.

Soesthetic group has completed over 30 projects both public and private.

The key of their success is their attention to detail and the high appreciation of their clients. Their clients especially value their ability to create new and innovative designs suited to their particular needs, preferences and lifestyles.

They supervise their projects from the very first sketch, until the last piece of furniture is in place. Throughout the entire building process they solve prolems on the spot, paying special attention to timing, materials selection and every detail.

They aspire to acheive maximum artistic expression in their works, to combine state-of-the-art technology with natural and eco-friendly materials. Their deep understanding of ergonomics and smart integration of engineering equipment helps create comfortable spaces where form and function coexist in perfect balance.

They have been awarded numerous prizes at many design competitions including ukraine's " Best Project of the Year" (2010 to 2014).

Seventh Studio

Seventh Studio was founded in 2003. Seventh Studio provides services for the design of interiors and buildings, monitoring the implementation of the construction works. Most of the customers contact the firm by the recommendation of friends and colleges, because they are satisfied. Seventh Studio is a leader in Ukraine in the restaurant design. Seventh Studio has designed more than 100 different restaurants in Ukraine, fifteen in Moscow, several in London and now several projects in other countries. To make restaurant special, Seventh Studio has to pay a lot of attention to details and develop specific elements for each restaurant. Over the years of work, Seventh Studio has gained enough experience to design anywhere in the world. The chief of studio is Yuriy Yumashev, leading architect is Yuriy Cherebedov, leading designer is Dmitriy Zhukov.

Wirt Design Group

Wirt Design Group

Wirt Design Group is a full-service commercial interior design firm located in Los Angeles, CA.

Founded in 1994, the firm has grown into one of Los Angeles' leading commercial interior design practices with a significant portfolio of projects for Clients such as Red Bull, Sempra Energy, Whole Foods Market, eHarmony, Yahoo!, and Northwestern Mutual. To this day, the founding principles of listening to and understanding each client's unique needs still underscore our approach to every project we undertake.

The hallmark of our practice is the cultivation of long-term client relationships and numerous clients have been with WDG for well over a decade. This history and loyalty speaks to our collective ability to listen to our clients and synthesize their needs and goals into an efficient and well-designed workplace.

Wirt Design Group ranks among the city's top commercial design practices and within the top 150 design firms in the United States.

Giant Leap

One Leap at a time, Giant Leap began in 2001 and has now evolved into the largest interior architecture-workplace specialist practice in Southern Africa comprising of more than 100 passionate staff and employees.

The company focuses exclusively on commercial projects with over 100 projects completed per annum. The professional teams consist of project managers, accounts executives, space planners, designers, quantity surveyors, architects and research teams, with experience in full turn key projects.

As a member of the Green Building Council, Giant Leap continually recommends products in line with the environmental regulations. Giant Leap's research division, Know More, can also assist with interior green ratings, delivering interior solutions that consider energy efficiency and recyclable products.

Giant Leap is driven by a simple idea, that if you treat people well, you bring out the very best in them. This has been the driving force behind their success over the years. It is built into every interior they design, because "when work is a pleasure, life is a joy!"

It's all about creating superior interiors that people love.

Superior Interiors inspire efficient and innovative work. They boost awareness of company values and allow people to experience a brand from the inside. These are the type of spaces that attract great clients and top talent. No matter the purpose or rhythm of the space, superior interiors are about feeling inspired because, when people are happier, business is better.

FORM

FORM Bureau

FORM is a multidisciplinary architecture and design practice established by Vera Odyn and Olga Treivas in Moscow in 2011.

The practice takes on projects of different scale, from minor temporary constructions to public spaces. Projects include the Garage Museum headquarters, the Garage Museum education center, design proposal for the National Gallery of Contemporary Art. Recently FORM won an international competition for Mitino Park in Moscow.

One of the directions of the practice is exhibition design. FORM has collaborated with major contemporary art institutions, working on exhibitions such as "5th Moscow Biennale of Contemporary Art", "Marina Abramovich: the Artist is present", "Dali & Media" and "Warhol's 10" among others.

SPARK

SPARK is an award-winning international design studio that creates distinctive buildings for clients and great places for people. SPARK focuses on architecture's potential to contribute positively to the experience of the city while addressing the pragmatic issues that govern each project. SPARK works with the bold yet common-sense vision of enlarging the spaces of the city into buildings, and of unfolding their buildings into the city-creating opportunities for layered experiences and engaging places.

SPARK's celebrated designs emerge from a detailed analysis of context, brief, and typology. SPARK has a multinational team numbering over 100. SPARK works synergistically, fostering numerous perspectives on culture and varied professional experience to achieve rich, integrated design solutions that consider the impact on all project stakeholders. From SPARK's four offices-in London, Beijing, Shanghai, and Singapore – SPARK has created and delivered projects in Asia, Europe, the Middle East, Africa, and Australia. Each one has its own unique spark and manifests desire to tackle and deliver on challenges that reflect the key global imperative of attaining a sustainable, life-improving environment for all.

SPARK's award winning projects include Clarke Quay in Singapore, the Shanghai International Cruise Terminal (MIPIM Asia Awards 2011, "Best Mixed-Use development" award), the Starhill Gallery Kuala Lumpur and the Raffles City projects in Ningbo and Beijing.

JUMP STUDIOS

Jump Studios

Jump Studios is a London based architecture and design practice. Established in 2001 by Shaun Fernandes and Simon Jordan, Jump Studios has completed several award-winning projects to date for clients including Nike, Google, Mother London, Levi Strauss, Starwood Hotels, Innocent Smoothies, Red Bull, Adidas, Wieden + Kennedy, The Science Museum and Honda. Jump is currently working on projects for Saatchi & Saatchi, Google, Yahoo! Waze and Rapha among others.

Studioninedots

Studioninedots

Studioninedots was established by Albert Herder, Vincent van der Klei, Arie van der Neut and Metin van Zijl in 2010 in Amsterdam. It is an international operating office with an architectural playing field ranging from designing a chair to developing a city. Studioninedots sees architecture as a reflection of contemporary urban culture which touches upon trends, movements, cultures, conflict, diversity, change, interaction, surprise, opportunities, threats, density, ecology and complexity. In its architecture and urban design the office integrates urban processes to get resourceful answers to current design queries.

The correlation between knowledge and creativity is the strength of the office. The inception of the design process is uncovering the essence of assignment. By making the project tangible, space is created for new ways of thinking, for influences outside architecture or urbanism, and to think and work interdisciplinary. Due to extensive experience the office is able to flawlessly get progressive ideas and designs into a realistic framework, in close collaboration with the client. Thus complex designs can be fast and effectively turned into reality.

Studioninedots designs sophisticated projects that define themselves through an apparent simplicity and a richness of experience. It creates buildings that express a constant sensory perception between recognition and surprise. By re-arranging the common-placed urban elements, buildings and plans are created that embody urban processes, giving them an agility to move with the dynamics of the city and therefore stay meaningful in a constant changing environment.

Won Myoung Hyun

President of INEX Design

Gachon University of Adjunct Professor of Interior Architecture

Kukmin University Graduate School of Design Interior Design major

2007 Korea Desin Award Space Designer 20 – Design House

2008 Modern Decoration International Media Prize

2009 Korea Interior Designer 100 – KOSID

2012 Next Interior Designer 30 – Interiors

2012 Korea Golden Scale Award

2014 Korea Golden Scale Award

2014 Best Interior Designer – Gain Design Group

Dorte Mandrup
Arkitekter A/S

Dorte Mandrup Arkitekter A/S (public limited company) is an international practice, based in Copenhagen, Denmark; founded in (1999), owned and managed by Dorte Mandrup, CEO, Professor & Architect MAA.

Oversight provided by a professional board of directors: Chairman Flemming Andersen, Line Rix, Peter Luke and Dorte Mandrup.

The studio is internationally renowned for original architectural design of high standards, from the overall scheme to precise detailing.

The office has developed expertise in creating playful, innovative and poetic solutions out of complex programmes within difficult contexts. This had led to acclaimed buildings, frequently used all over the world as references for visionary solutions.

The work process is based on an open dialogue with the client and a thorough understanding of a project's demands and possibilities, ensuring a holistic answer to each challenge.

The portfolio, highlighting the firm's flexibility and versatility, includes cultural institutions, buildings for children and youth, sports facilities, schools, housing, master plans and offices, as well as renovation and alteration of historically significant buildings.

The organization of the office allows founder and sole-owner Dorte Mandrup to work "hands on" and in close interaction with all project teams, bringing careful guidance and high quality to each project.

Dorte's experience, judgment and ability to see the right solution is frequently recognized and used in international jury contexts, and in her work as a lecturer and professor.

rockwellgroup

Rockwell Group

With a desire to create immersive environments, Rockwell Group takes a cross-disciplinary approach to its inventive array of projects. Based in downtown New York and with satellite offices in Madrid and Shanghai, our innovative, internationally acclaimed architecture and design firm specializes in hospitality, cultural, healthcare, educational, product, theater, and film design. Crafting a unique and individual narrative concept for each project is fundamental to Rockwell Group's successful design approach. From the big picture to the last detail, the story informs and drives the design. The seamless synergy of technology, craftsmanship, and design is reflected in environments that combine high-end video technology, handmade objects, special effects, and custom fixtures and furniture.

Honors include the 2008 National Design Award from the Cooper-Hewitt National Design Museum for outstanding achievement in Interior Design; the 2009 Pratt Legends Award; the Presidential Design Award for his renovation of the Grand Central Terminal; induction into the James Beard Foundation Who's Who of Food & Beverage in America and Interior Design magazine's Hall of Fame; inclusion in Architectural Digest's AD 100; three Tony Award nominations for Best Scenic Design; and four Drama Desk Award nominations for Outstanding Scenic Design of a Musical. Rockwell Group was named by Fast Company as one of the most innovative design practices in their annual World's 50 Most Innovative Companies issue in 2009, 2014 and 2015.

Joliark

Joliark is an architectural practice based in Stockholm. The main focus is developing and running projects within the fields of urban planning, house construction and interior design with the ambition to heighten the architectural qualities of the physical environment. Joliark is owned by Per Johanson, Helen Johansson, Hans Linnman, Magnus Pörner and Cornelia Thelander. The practice was founded in 1972 following an awarded first prize in a Nordic architectural competition for the design of Vanda Centrum outside Helsinki.

Arquitectura e Interiores SAS

During 19 years Arquitectura e Interiores SAS has designed and built more than 1.5 million square meters and has certified the highest number of LEED projects in Colombia. The company has created spaces to foster productivity, team work and sense of belonging. All projects are the result of a dedicated and detailed effort, with a deep understanding of business and industry. The company has been leader in Colombia in terms of experience and innovation in each design. Global companies such as P&G, Unilever, GlaxoSmithKline, Pfizer, IBM, Coca Cola, among others have trusted their corporate spaces in Colombia to this firm.

Estudio Guto Requena

Estudio Guto Requena reflects about memory, digital culture and poetic narratives in all design scales. Guto, 35 years old, was born in Sorocaba, countryside of São Paulo State. He graduated as Architect and Urban Planner in 2003 at University of São Paulo. During nine years he was a researcher at NOMADS, center for Interactive Living Studies of the University of São Paulo. In 2007 he got his Master degree at the same University with the dissertation, Hybrid Habitation: Interactivity and Experience in the Cyberculture Era.

He won awards and had lectured and exhibited in several cities worldwide as New York, Milan, Paris, Istanbul, Moscow, Dubai, Mexico City, Santiago, Cape Town, Beijing, Bangkok, Miami and London. He was a professor at Panamericana – School of Arts and Design and at IED – Istituto Europeo di Design – at both graduation and master levels. Guto had lectured on 70 workshops all over the country, and received the Young Brazilian Awards recognition. In 2012 Guto was selected by Google to develop the project for their Brazilian headquarter. And in 2013 Walmart selected him to design their headquarter, wining the international award "Building of the Year" from Archdaily with this project under "Interior Architecture" Category.

Since 2012 Guto has a column at newspaper Folha de São Paulo where he writes about design, architecture and urbanism and collaborates writing for many magazines. In 2011 Guto created, wrote and hosted the TV show Nos Trinques, for Brazilian TV Globo channel GNT and developed design web series for the same channel, recorded in Milan, Paris, Amsterdam and London.

Ma Hui

With more than ten years in the interior design, Mr. Ma not only can precisely meet customers needs, but also blends the concept of "design creates value" into every project and forms a unique design style. Moreover, he has presided over the design of Hangzhou International Airport, Qingdao Liuting International Airport, Zhejiang TV and other big public projects and becomes an influential designer representative in the industry.

In recent years, Ma Hui and his company have focused on researching real estate product design and providing a full set of professional services. He has cooperated with Greentown, Evergrande, Gemdale, Poly, Sino-Ocean and other well-known real estate companies, has won many awards in the interior design competition and reported and interviewed by professional magazines and media for several times. As a renowned designer in the industry, Mr. Ma has been selected as the annual cover Chinese interior designer, the outstanding designer in the 20th anniversary celebration of the Interior Design Branch of CIID Architectural Society of China, one of the Top Ten Annual Designer winning Golden Goat Awards in Guangdong International Design Week, etc. and awarded the China Interior Design Elite Award. Moreover, he has been interviewed by many renowned network media, newspapers, magazines and other well-known media.

ACKNOWLEDGEMENT 鸣谢

We would like to thank everyone involved in the production of this book, especially all the artists, designers, architects and photographers, for their kind permission to publish their works, for their contribution of images, ideas and concepts, as well as their creativity to be shared with readers around the world.

  在此,我们非常感谢参与本书编写的所有人员,尤其是各位艺术家、设计师、建筑师和摄影师,他们授权我们出版他们的作品,感谢他们提供的图片、思想和理念,并将他们的创意分享给全世界的读者。